# The
# International
# Menu Speller

### Kenneth N. Anderson
### and
### Lois E. Anderson

**John Wiley & Sons, Inc.**
New York • Chichester • Brisbane • Toronto • Singapore

*This text was printed on acid-free paper.*

This publication is designed to provide accurate and authoritative information in regard to the subject matter covered. It is sold with the understanding that the publisher is not engaged in rendering legal, accounting, or other professional service. If legal advice or other expert assistance is required, the services of a competent professional person should be sought. FROM A DECLARATION OF PRINCIPLES JOINTLY ADOPTED BY A COMMITTEE OF THE AMERICAN BAR ASSOCIATION AND A COMMITTEE OF PUBLISHERS.

Library of Congress Cataloging-in-Publication Data

Anderson, Kenneth, 1921–
    The international menu speller / by Kenneth N. Anderson and Lois E. Anderson.
        p.   cm.
    Includes index.
    ISBN 0-471-58435-5 (alk. paper)
    1. Cookery, International—Terminology.   2. Food—Terminology.
3. Nutrition—Terminology.   4. Language and languages—Orthography and spelling—Dictionaries.   I. Anderson, Lois E.   II. Title.
TX725.A1A56   1993
641.59'03—dc20                                                          93-7261
                                                                        CIP

Printed in the United States of America

10   9   8   7   6   5   4   3   2   1

# Preface

This convenient little reference book has been specially designed for foodservice professionals and others who create menus and write about food. It will help you

- Create a professional, polished menu, the all-important marketing tool in today's competitive industry.
- Avoid spelling errors in writing descriptions of foods and food preparations.

With over 7,500 listings, this book goes beyond the basic culinary vocabulary. Based on *The International Dictionary of Food and Nutrition*, it includes food words in over 40 languages—including the sometimes confusing words associated with ethnic and regional cooking as well as the nutritional concepts that interest today's health-minded foodservice customers.

You will find culinary and nutrition terms, names of ingredients, sauces, seasonings, preparation techniques, and other words needed in writing menus.

THE INTERNATIONAL MENU SPELLER is easy to use. For quick reference, all of the entries are presented in one alphabetical list. Each word is presented first as it looks when printed properly, then hyphenated for proper word breaks. A country code indicates the word's original language (see the abbreviation key following).

The following is a list of the abbreviations used to indicate the more than forty languages used in this book. In some cases, an abbreviation is modified to further clarify. For example, (Gr–Swiss) means German spoken in Switzerland or (Af–Swahili) means the Swahili spoken in vast areas of Africa.

| | | | |
|---|---|---|---|
| (Aa) | Australian | (Jw) | Jewish (includes American-Jewish and Israeli terms) |
| (Af) | African | | |
| (Ar) | Arabic | | |
| (Bl) | Belgian | (Kr) | Korean |
| (Bu) | Bulgarian | (Ml) | Malaysian |
| (Ca) | Canadian | (Mx) | Mexican |
| (Cb) | Caribbean | (Nw) | Norwegian |
| (Ch) | Chinese | (Pg) | Portuguese (includes Portugal and Brazil) |
| (Cz) | Czechoslovakian | | |
| (Da) | Danish | | |
| (Du) | Dutch | (Ph) | Philippine |
| (Fi) | Finnish | (Pl) | Polynesian (includes Hawaii) |
| (Fr) | French | | |
| (GB) | British (includes England and Wales) | (Po) | Polish |
| | | (Ro) | Romanian |
| | | (Rs) | Russian |
| (Gk) | Greek | (Sc) | Scottish |
| (Gr) | German | (SC) | Serbo-Croatian |
| (Hu) | Hungarian | (Sp) | Spanish |
| (Ia) | Hindi (India) | (Sw) | Swedish |
| (In) | Indonesian | (Th) | Thai |
| (Ir) | Irish | (Tr) | Turkish |
| (It) | Italian | (US) | American |
| (Jp) | Japanese | (Vt) | Vietnamese |

# *A*

***aab ghosht*** (Ia), aab ghosht
***Aal*** (Gr, Du), Aal
***Aalbutte*** (Gr), Aal-but-te
***Aalgeräuchert*** (Gr), Aal-ge-räu-chert
***aalgestoofd*** (Du), aal-ge-stoofd
***Aalgrün*** (Gr), Aal-grün
***aal i gélé*** (Da), aal i gé-lé
***aalmutter*** (Gr), aal-mut-ter
***aalsoep*** (Du), aal-soep
***Aalsuppe*** (Gr), Aal-sup-pe
***aam*** (Ia), aam
***a'ama*** (Pl), a'am-a
***aam chatni*** (Ia), aam chat-ni
***aamiainen*** (Fi), aa-mi-a-i-nen
***aamiaisherkku*** (Fi), aa-mi-a-is-herk-ku
***aamiaispaistos*** (Fi), aa-mi-a-is-pa-is-tos
***aamiaissämpylät*** (Fi), aa-mi-a-is-säm-py-lät
***aardappel*** (Du), aar-dap-pel
***aardappelsoep*** (Du), aar-dap-pel-soep
***aardbei-chipolata*** (Du), aard-bei-chi-po-la-ta
***aardbeien*** (Du), aard-bei-en
***aardvark*** (Af), aard-vark
***aardwolf*** (Af), aard-wolf
***Aargauer Rüeblitorte*** (Gr), Aar-gau-er Rüe-bli-tor-te

***Aaron's rod*** (US), Aar-on's rod
***aassida*** (Af), a-as-si-da
***aata*** (Ia), aa-ta
***aawa*** (Pl), aa-wa
***abacate*** (Pg), a-ba-ca-te
***abacate batida*** (Pg), a-ba-ca-te ba-ti-da
***abacaxi*** (Pg-Brazil), a-ba-ca-xi
***abadejo*** (Sp), a-ba-de-jo
***abaisse*** (Fr), a-baisse
***abalone*** (US), ab-a-lo-ne
***abats*** (Fr), a-bats
***abats d'agneau*** (Fr), a-bats d'agn-eau
***abats de boucherie*** (Fr), a-bats de bou-cher-ie
***abattis*** (Fr), a-bat-tis
***abattis de volaille*** (Fr), a-bat-tis de vo-laille
***abattis en ragoût*** (Fr), a-bat-tis en ra-goût
***abba*** (Ia), ab-ba
***abbacchi arrosto*** (It), ab-bac-chi ar-ros-to
***abbacchio*** (It), ab-bac-chio
***abbacchio al forno*** (It), ab-bac-chio al for-no
***abboccato*** (It), ab-boc-ca-to
***abbor*** (Nw), ab-bor
***abbrustolire*** (It), ab-brus-to-lire

**abdug** (Ar), ab-dug
**abelmosk** (US), a-bel-mosk
**Abendessen** (Gr), A-ben-dessen
**Aberdeen Angus** (Sc), Ab-erdeen An-gus
**Aberdeen roll** (Sc), Ab-erdeen roll
**Abernethy biscuit** (Sc), Ab-er-neth-y bis-cuit
**Abertam** (Cz), A-ber-tam
**abfetten** (Gr), ab-fet-ten
**abfüllen** (Gr), ab-fül-len
**abijau** (Ar), a-bi-jau
**abkochen** (Gr), ab-ko-chen
**abkochen Milch** (Gr), ab-ko-chen Milch
**à blanc** (Fr), à blanc
**ablette** (Fr), ab-lette
**abóbora** (Pg), a-bó-bo-ra
**aboloo** (Af), a-bo-loo
**abon ayam** (In), a-bon a-yam
**aborinha** (Pg), a-bor-i-nha
**aboukir** (Ar), a-bou-kir
**a brasileira** (Pg), a bras-i-le-i-ra
**abricot** (Fr), ab-ri-cot
**abricot-péche** (Fr), ab-ri-cot-péche
**abricots Condé** (Fr), ab-ri-cots Con-dé
**abrikos** (Rs), a-bri-kos
**abrikosovi povidlo** (Rs), a-bri-kos-o-vi po-vid-lo
**abrikosovi sup** (Rs), a-bri-kos-o-vi sup
**abrikossnitte** (Da), a-bri-kos-snit-te
**abuñolado** (Sp), a-bu-ño-la-do

**abura** (Jp), a-bu-ra
**abura age** (Jp), a-bu-ra a-ge
**aburakkokunai** (Jp), a-bu-rak-ko-ku-nai
**abura miso** (Jp), a-bu-ra mi-so
**abutilon** (Pg), a-bu-til-on
**acacia** (US), a-ca-cia
**Acadian blueberry grunt** (Ca), A-ca-di-an blue-ber-ry grunt
**Acadian cheese bread** (Ca), A-ca-di-an cheese bread
**açafrão** (Pg), a-ça-frã-o
**acajou** (Fr), a-ca-jou
**acajú** (Pg), a-ca-jú
**acara** (Pg), a-ca-ra
**acaramelado** (Pg), a-car-a-me-la-do
**acar campur** (In), a-car cam-pur
**acar ikan** (In), a-car i-kan
**acarne** (Fr), a-carne
**Ac'cent** (US), Ac'cent
**acciughe** (It), ac-ciu-ghe
**accra de morue** (Cb), ac-cra de mo-rue
**acecinado** (Sp), a-ce-cin-a-do
**acedera** (Sp), a-ce-der-a
**aceite** (Sp), a-cei-te
**aceitunas** (Sp), a-cei-tu-nas
**aceitunas negras** (Sp), a-cei-tu-nas ne-gras
**acelga** (Sp), a-cel-ga
**acelgas con crema** (Sp), a-cel-gas con cre-ma
**acerola** (Cb), ac-er-o-la
**acesulfame-K** (GB), ac-e-sul-fame-K

*aceteria* (Sp), a-ce-te-ri-a
*acetic acid* (US), ac-e-tic ac-id
*acetini* (It), a-ce-ti-ni
*aceto* (It), a-ce-to
*aceto balsamico di Modeno* (It), a-ce-to bal-sa-mi-co di Mo-de-no
*aceto-dolce* (It), a-ce-to-dol-ce
*acétomel* (Fr), a-cé-to-mel
*acétoselle* (Fr), a-cé-to-selle
*achar* (Ia), a-char
*achara zuke* (Jp), a-cha-ra zu-ke
*achari* (Af-Swahili), a-cha-ri
*achar tandal* (Ia), a-char tan-dal
*ache* (Fr), ache
*achicoria* (Sp), a-chi-cor-ia
*achigan* (Fr), ach-i-gan
*achiote* (Sp), a-chi-o-te
*acidophilus milk* (US), ac-i-doph-i-lus milk
*acids* (US), ac-ids
*acidulated water* (US), a-ci-du-lat-ed wa-ter
*acini di pepe* (It), a-ci-ni di pe-pe
*ackee* (Cb), ac-kee
*açorda* (Pg), a-çor-da
*açorda alentejana* (Pg), a-çor-da a-len-te-ja-na
*açorda de alho* (Pg), a-çor-da de a-lho
*acorn* (US), a-corn
*acorn squash* (US), a-corn squash
*acqua* (It), ac-qua
*acquacotta* (It), ac-qua-cot-ta
*acqua minerale* (It), ac-qua mi-ne-ra-le
*acrid* (US), ac-rid
*acrolein* (US), ac-ro-le-in
*active dry yeast* (US), ac-tive dry yeast
*açúcar* (Pg), a-çú-car
*adafina* (Sp), a-da-fi-na
*additive* (US), ad-di-tive
*ådelost* (Sw), å-del-ost
*aderezo de comida* (Sp), a-de-re-zo de co-mi-da
*adobo* (Ph), a-do-bo
*adobo criollo* (Sp), a-do-bo cri-o-llo
*adobo de pescado* (Sp), a-do-bo de pes-ca-do
*adobong labong* (Ph), a-do-bong la-bong
*adobong pusit* (Ph), a-do-bong pu-sit
*adobo sauce* (Mx), a-do-bo sauce
*adrak* (Ia), ad-rak
*adrak chatni* (Ia), ad-rak chat-ni
*adrak murgh* (Ia), ad-rak murgh
*adriatico, dell'* (It), ad-ri-a-ti-co, dell'
*advocaat* (Du), ad-vo-caat
*adzhersandal* (Rs), ad-zher-san-dal
*adzuki* (Jp), ad-zu-ki
*æbleflæsk* (Da), æ-ble-flæsk
*æblegrød* (Da), æ-ble-grød
*æblekage* (Da), æ-ble-ka-ge
*æbleskiver* (Da), æ-ble-ski-ver
*æg* (Da), æg

**æggekage** (Da), æg-ge-ka-ge
**æg og sild** (Da), æg og sild
**aemono** (Jp), ae-mo-no
**ærter** (Da), ær-ter
**affettato** (It), af-fet-ta-to
**affogato** (It), af-fo-ga-to
**affumicato** (It), af-fu-mi-ca-to
**agachadiza** (Sp), ag-a-cha-di-za
**agar-agar** (Ml), a-gar-a-gar
**agedashi** (Jp), ag-e-da-shi
**agemono** (Jp), ag-e-mo-no
**ageta** (Jp), ag-e-ta
**age, to** (US), age, to
**ägg** (Sw), ägg
**äggröra** (Sw), ägg-rö-ra
**aglio** (It), ag-lio
**aglio e olio** (It), ag-lio e o-lio
**agnautka** (Rs), ag-naut-ka
**agneau** (Fr), agn-eau
**agneau de lait persillé** (Fr), agn-eau de lait per-sillé
**agneau grillé au thym** (Fr), agn-eau gril-lé au thym
**agnello** (It), a-gnel-lo
**agnello all' arrabbuata** (It), a-gnel-lo al-l' ar-rab-bu-a-ta
**agneshko magdanoslija** (Bu), ag-nesh-ko mag-da-nos-li-ja
**agnolotti** (It), a-gno-lot-ti
**agnolotti di grasso** (It), a-gno-lot-ti di gras-so
**agoni** (It), a-gon-i
**agoni seccati in graticola** (It), a-gon-i sec-ca-ti in gra-ti-co-la
**agourelo** (Gk), a-gou-re-lo
**agresto** (It), a-gres-to

**agrião** (Pg), a-gri-ã-o
**agro** (Sp), a-gro
**agrodolce** (It), a-gro-dol-ce
**agua** (Sp), a-gua
**água** (Pg), á-gua
**aguacate** (Sp), a-gua-ca-te
**aguacate encamaronados** (Mx), a-gua-ca-te en-cam-a-ron-a-dos
**aguacate picante** (Sp), a-gua-ca-te pi-can-te
**aguacates rellenos** (Sp), a-gua-ca-tes rel-le-nos
**água com gelo** (Pg), á-gua com ge-lo
**água mineral** (Pg), á-gua mi-ne-ral
**agurk** (Da), a-gurk
**agurkai su rukcscia grietine** (Rs), a-gurk-ai su ruk-cscia gri-e-ti-ne
**agurker** (Nw), a-gur-ker
**agurkesalat** (Da), a-gur-ke-sa-lat
**ahds** (Ar), ahds
**ahds imqala** (Ar), ahds im-qa-la
**ahds imsafa** (Ar), ahds im-sa-fa
**ahds majroosh** (Ar), ahds maj-roosh
**ahds polo** (Ar), ahds po-lo
**ahi** (Pl), a-hi
**ahjeen** (Ar), ah-jeen
**ahjeen il fatayer** (Ar), ah-jeen il fa-tay-er
**ahmeeghthalota** (Gk), ah-meegh-tha-lo-ta
**ahngooree** (Gk), ahn-goo-ree
**ahsal** (Ar), ah-sal

**abududu** (Tr), a-hu-du-du
**abven** (Fi), ah-ven
**ai ferri** (It), a-i fer-ri
**aiglefin** (Fr), ai-gle-fin
**aïgo sau d'iou** (Fr), aï-go sau
  d'iou
**aigre-doux** (Fr), ai-gre-doux
**aiguillette** (Fr), ai-gui-llette
**aiguillette de canetons
  Montmorency** (Fr), ai-gui-
  llette de can-e-tons Mont-
  mor-en-cy
**ail** (Fr), ail
**aile de poulet** (Fr), aile de
  pou-let
**aillade** (Fr), ai-lla-de
**aïoli** (Fr), aï-o-li
**aipo** (Pg), a-i-po
**air** (In), a-ir
**airelle** (Fr), ai-relle
**air jeruk manis** (In), a-ir je-
  ruk ma-nis
**air tomat** (In), a-ir to-mat
**aish** (Ar), aish
**aisu kōhi** (Jp), a-isu kō-hī
**aisu kurīmu** (Jp), a-isu kurīmu
**aisu tī** (Jp), a-isu tī
**ajawn seeds** (Ia), a-jawn
  seeds
**aji** (Jp), a-ji
**aji** (Sp), a-ji
**aji-no-moto** (Jp), a-ji-no-mo-
  to
**ajmoda** (Ia), aj-mo-da
**ajo** (Sp), a-jo
**ajo cebollino** (Sp), a-jo ce-
  bol-li-no
**ajókamártas** (Hu), a-jó-ka-
  már-tas

**ajonjoli** (Sp), a-jon-jo-li
**ajo porro** (Sp), a-jo por-ro
**akadashi** (Jp), a-ka-da-shi
**aka miso** (Jp), a-ka mi-so
**akee** (Cb), a-kee
**åkerbøne** (Nw), å-ker-hø-ne
**akevitt** (Nw), a-ke-vitt
**akhinos** (Gk), akh-i-nos
**akhladhi** (Gk), akh-la-dhi
**akhrot** (Ia), akh-rot
**akkra** (Cb), ak-kra
**akuri** (Ia), a-ku-ri
**akvavit** (Da, Sw), ak-va-vit
**al** (It), al
**ål** (Da, Nw, Sw), ål
**à la, à l'** (Fr), à la, à l'
**a la** (Sp), a la
**alabalik** (Tr), a-la-ba-lik
**alabega** (Sp), a-la-be-ga
**à la carte** (Fr), à la carte
**alajú** (Sp), a-la-jú
**à l'algérienne** (Fr), à l'al-gér-
  ienne
**à l'alsacienne** (Fr), à l'al-sa-
  cienne
**à l'amiral** (Fr), à l'am-i-ral
**à la mode** (Fr), à la mode
**à la mode de** (Fr), à la mode
  de
**à l'ancienne** (Fr), à l'an-
  cienne
**à l'andalouse** (Fr), à l'an-da-
  louse
**à l'anglaise** (Fr), à l'an-glaise
**al arancio** (It), al a-ran-cio
**à l'Argenteuil** (Fr), à l'Ar-gen-
  teuil
**à l'arlésienne** (Fr), à l'arl-és-
  ienne

*alaskačorba* (SC), a-la-ska-
čor-ba

*à l'autrichienne* (Fr), à l'au-
tri-chienne

*alb* (Ar), alb

*albacore* (US), al-ba-core

*albaricoque* (Sp), al-bar-i-
coque

*Albert, sauce* (Fr), Al-bert,
sauce

*albicocca* (It), al-bi-coc-ca

*albillo* (Sp), al-bi-llo

*albóndigas* (Sp), al-bón-di-
gas

*Albuféra, sauce* (Fr), Al-bu-
fé-ra, sauce

*albumin* (US), al-bu-min

*al burro* (It), al bur-ro

*alcachofas a la vinagreta*
(Sp), al-ca-cho-fas a la vi-
na-gre-ta

*alcachofra* (Pg), al-ca-cho-fra

*alcaparras* (Pg), al-ca-par-ras

*alcaravea* (Sp), al-car-a-ve-a

*alcobol* (US), al-co-hol

*alcool* (Fr, Sp), al-cool

*al dente* (It), al den-te

*ale* (US), ale

*ålesuppe* (Da), ål-e-sup-pe

*alewife* (US), ale-wife

*alface* (Pg), al-fa-ce

*alfajor* (Sp), al-fa-jor

*alfalfa sprouts* (US), al-fal-fa
sprouts

*al forno* (It), al for-no

*ålgstek* (Sw), ålg-stek

*albo* (Pg), a-lho

*albo-poró* (Pg-Brazil), a-lho-
po-ró

*al borno* (Sp), al hor-no

*ali* (It), a-li

*alice* (It), al-i-ce

*alici sott'olio* (It), a-li-ci sot-
t'o-lio

*alicot* (Fr), a-li-cot

*aliñado* (Sp), a-li-ña-do

*aliño* (Sp), a-li-ño

*ali-oli* (Sp), a-li-o-li

*alkali* (US), al-ka-li

*alkanet* (US), al-ka-net

*Alkobol* (Gr), Al-ko-hol

*alkupalat* (Fi), al-ku-pa-lat

*alla, alle, allo* (It), al-la, al-le,
al-lo

*alla brace* (It), al-la bra-ce

*all'agliata* (It), al-l'agl-ia-ta

*all'amatriciana* (It), al-l'a-
ma-tri-cia-na

*alle acciugbe* (It), al-le ac-
ciu-ghe

*alle cozze* (It), al-le coz-ze

*allemande, sauce* (Fr), al-le-
mande, sauce

*alle vongole* (It), al-le von-
go-le

*Allgäuer Emmentbaler* (Gr),
All-gäu-er Emmenthaler

*Allgewürz* (Gr), All-ge-würz

*alligator pear* (US), al-li-ga-
tor pear

*all'olio ed aglio* (It), al-l'o-lio
ed agl-io

*alloro* (It), al-lo-ro

*allo spiedo* (It), al-lo spi-e-do

*allspice* (Cb), all-spice

*allumettes* (Fr), all-u-mettes

*all'uovo* (It), al-l'uo-vo

*alma* (Hu), al-ma

*almamártás* (Hu), al-ma-
már-tás

**almás rétes** (Hu), al-más ré-
tes
**almejas** (Sp), al-me-jas
**almejas en salsa de ajo** (Sp),
al-me-jas en sal-sa de a-jo
**almendra** (Sp), al-men-dra
**almendra amarga** (Sp), al-
men-dra a-mar-ga
**almendrado** (Sp), al-men-
dra-do
**almendras confitadas** (Sp),
al-men-dras con-fi-ta-das
**almendras de cacao** (Sp),
al-men-dras de ca-ca-o
**almendras garapiñadas**
(Sp), al-men-dras ga-ra-pi-
ña-das
**almendras tostados** (Sp), al-
men-dras tos-ta-dos
**almibar** (Sp), al-mi-bar
**almibares** (Sp), al-mí-bar-es
**almirón** (Sp), al-mi-rón
**almodón** (Sp), al-mo-dón
**almodrote** (Sp), al-mo-dro-te
**almojábana** (Sp), al-mo-já-
ba-na
**almond** (US), al-mond
**almôndegas** (Pg), al-môn-
de-gas
**almori** (Sp), al-mo-ri
**almorzar** (Sp), al-mor-zar
**ål og røræg** (Da), ål og rør-
æg
**aloque** (Sp), a-lo-que
**alosa** (Sp), a-lo-sa
**alose** (Fr), a-lo-se
**alouettes** (Fr), al-ouettes
**alouettes sans têtes** (Fr), al-
ouettes sans têtes
**aloyau** (Fr), al-o-yau

**alperche** (Pg), al-per-che
**alpistela** (Sp), al-pi-ste-la
**Alpkäse** (Gr-Swiss), Alp-kä-se
**al ragù** (It), al ra-gù
**al sangue** (It), al san-gue
**Alse** (Gr), Al-se
**al sugo** (It), al su-go
**alu** (Ia), a-lu
**alubia** (Sp), a-lu-bia
**alu bukhara** (Ia), a-lu bu-
kha-ra
**alu chat** (Ia), a-lu chat
**alu matar** (Ia), a-lu ma-tar
**alu pakoras** (Ia), a-lu pa-ko-
ras
**amai** (Jp), a-ma-i
**amande** (Fr), am-ande
**amande amère** (Fr), am-
ande a-mère
**amandel** (Du), a-man-del
**amandine** (Fr), a-man-dine
**amaranth** (US), am-a-ranth
**amarelle** (US), am-a-relle
**amaretti** (It), a-ma-ret-ti
**amaretto** (It), a-ma-ret-to
**amaro** (It), a-ma-ro
**amazake** (Jp), a-ma-za-ke
**amazu** (Jp), a-ma-zu
**ambrosia** (It), am-bro-sia
**Ambrosia** (Sw), Am-bro-sia
**amchoor** (Ia), am-choor
**amêijoas** (Pg), a-mê-i-jo-as
**amêijoas na cataplana**
(Pg), a-mê-i-jo-as na ca-ta-
pla-na
**ameixas** (Pg), a-me-i-xas
**amêndoas** (Pg), a-mên-do-as
**amendoim** (Pg), a-men-do-
im

*américaine* (Fr), a-mé-ri-caine

*American cheese* (US), A-mer-i-can cheese

*amino acid* (US), a-min-o ac-id

*amirty* (Ia), a-mir-ty

*amóras* (Pg), a-mó-ras

*amrood* (Ia), am-rood

*anacard* (Pg-Brazil), a-na-card

*anadama bread* (US), an-a-dam-a bread

*anago* (Jp), a-na-go

*Anaheim chili* (US), An-a-heim chil-i

*ananas* (Cz, Fi, Nw, Rs, Tr), a-na-nas

*ananas au kirsch* (Fr), a-na-nas au kirsch

*ananasový meloun* (Cz), a-na-na-so-vý me-lo-un

*ananasso* (It), a-na-nas-so

*ananász* (Hu), a-na-nász

*anar* (Ia), a-nar

*anchellini* (It), an-chel-li-ni

*ancho* (Mx), an-cho

*anchoas* (Sp), an-choas

*anchois* (Fr), an-chois

*anchoussi s yaitzami* (Rs), an-chous-si s ya-it-za-mi

*anchouwa* (Ar), an-chou-wa

*anchova* (Pg), an-cho-va

*anchovies* (US), an-cho-vies

*anchoyade* (Fr), an-cho-yade

*ançuvez* (Tr), an-çu-vez

*and* (Da, Nw), and

*anda* (Ia), an-da

*anda ki kari* (Ia), an-da ki ka-ri

*andalouse, sauce* (Fr), an-da-louse, sauce

*andesteg* (Da), an-de-steg

*andijvie* (Du), an-dij-vie

*andouille* (Fr), and-ouille

*andruty* (Po), an-dru-ty

*anellini* (It), a-nel-li-ni

*anequim* (Pg), a-ne-quim

*aneth* (Fr), a-neth

*aneto* (It), a-ne-to

*angel food cake* (US), an-gel food cake

*angel hair* (US), an-gel hair

*angelica* (US), an-gel-i-ca

*angélique* (Fr), an-gé-lique

*anglerfish* (US), ang-ler-fish

*Anglesey eggs* (GB), Ang-le-sey eggs

*angoor* (Ia), an-goor

*angrešt* (Cz), an-grešt

*angsa* (In), ang-sa

*anguila* (Sp), an-gui-la

*anguilla marinata* (It), an-guil-la ma-ri-na-ta

*anguille* (Fr), ang-uille

*anguille alla veneziana* (It), ang-uil-le al-la ve-ne-zia-na

*anguilles au vert* (Bl), an-guilles au vert

*anguria* (It), an-gu-ri-a

*anice* (It), ani-ce

*anijs* (Du), a-nijs

*animelle* (It), a-ni-mel-le

*animelles à la crème* (Fr), an-i-melles à la crème

*anise* (US), an-ise

*anisetta* (It), an-i-set-ta

*anitra* (It), a-ni-tra

*anitra arrosto* (It), a-ni-tra ar-ro-sto

*anitra selvatica* (It), a-ni-tra
sel-va-ti-ca
*anjeer* (Ia), an-jeer
*Anjou* (Fr), An-jou
*anjova* (Sp), an-jo-va
*anjovis* (Fi), an-jo-vis
*anka* (Sw), an-ka
*ankerias* (Fi), an-ke-ri-as
*ankka* (Fi), ank-ka
*Annabella* (It), An-na-bel-la
*annatto* (US), an-nat-to
*anolini* (It), a-no-li-ni
*Anschovis* (Gr), An-scho-vis
*ansjovis* (Sw), an-sjo-vis
*antioxidants* (US), an-ti-ox-i-
dants
*antipasto* (It), an-ti-pa-sto
*antipasto variato* (It), an-ti-
pa-sto va-ria-to
*antojos* (Mx), an-to-jos
*aoyagi* (Jp), ao-ya-gi
*apams* (Ia), a-pams
*apel* (In), a-pel
*apelsin* (Sw), a-pel-sin
*apel'sin* (Rs), a-pel'sin
*apenoten* (Du), a-pe-no-ten
*apéritif* (Fr), a-pé-ri-tif
*aperitivo* (It), a-pe-ri-ti-vo
*Apfel* (Gr), Ap-fel
*Apfelkuchen* (Gr), Ap-fel-ku-
chen
*Apfelmus* (Gr), Ap-fel-mus
*Apfelpfannkuchen* (Gr), Ap-
fel-pfann-ku-chen
*Apfelreis* (Gr), Ap-fel-reis
*Apfelrotkohl* (Gr), Ap-fel-rot-
kohl
*Apfelsinen* (Gr), Ap-fel-si-nen
*Apfelsinensaft* (Gr), Ap-fel-
si-nen-saft

*Apfelstrudel* (Gr), Ap-fel-
stru-del
*Apfelwein* (Gr), Ap-fel-wein
*aphelia* (Gk), a-phe-lia
*aphrodisiac* (US), aph-ro-
dis-i-ac
*api'i* (Pl), a-pi'i
*apio* (Sp), a-pio
*apio-nabo* (Sp), a-pio-na-bo
*à point* (Fr), à point
*appelbeignets* (Du), ap-pel-
beign-ets
*äpplelkaka med vaniljsås*
(Sw), äp-ple-ka-ka med va-
nilj-sås
*appelmoes* (Du), ap-pel-
moes
*appelsap* (Du), ap-pel-sap
*appelsiini* (Fi), ap-pel-sii-ni
*appelsiinimehua* (Fi), ap-
pel-sii-ni-me-hua
*appelsin* (Nw), ap-pel-sin
*appelsinsaft* (Nw), ap-pel-
sin-saft
*appeltaart* (Du), ap-pel-taart
*Appenzell* (Gr) (Swiss), Ap-
pen-zell
*appetizer* (US), ap-pe-ti-zer
*apple* (US), ap-ple
*åpple* (Sw), åp-ple
*apple amber pudding* (GB),
ap-ple am-ber pud-ding
*apple butter* (US), ap-ple but-
ter
*apple charlotte* (GB), ap-ple
char-lotte
*apple dumplings* (GB), ap-
ple dump-lings
*applejack* (US), ap-ple-jack

**äppleknyten** (Sw), äp-ple-kny-ten

**apple pandowdy** (US), ap-ple pan-dow-dy

**apple pie** (US), ap-ple pie

**applesauce** (US), ap-ple-sauce

**apple slump** (US), ap-ple slump

**apple snow** (US), ap-ple snow

**appum** (Ia), ap-pum

**apricot** (US), a-pri-cot

**aprikoosi** (Fi), ap-ri-koo-si

**aprikos** (Nw, Sw), a-pri-kos

**Aprikosen** (Gr), Ap-ri-ko-sen

**aprósütemények** (Hu), ap-ró-sü-te-mé-nyek

**aquavit** (US), a-qua-vit

**arachide** (Fr, It), a-ra-chide

**arachis huile** (Fr), a-ra-chis huile

**aragosta** (It), a-ra-go-sta

**arak** (Ar), a-rak

**arancia** (It), a-ran-cia

**aranciata** (It), a-ran-cia-ta

**arancini** (It), a-ran-ci-ni

**arándano** (Sp), a-rán-da-no

**aranygaluska** (Hu), a-rany-ga-lus-ka

**arare** (Jp), a-ra-re

**arbei** (In), ar-bei

**arborio** (It), ar-bo-rio

**Arbroath smokies** (Sc), Ar-broath smok-ies

**arbuz** (Rs), ar-buz

**archiduc, à la** (Fr), ar-chi-duc, à la

**arenque** (Pg), a-ren-que

**arenque ahumado** (Sp), a-ren-que a-hu-ma-do

**arepas** (Sp), a-re-pas

**arhar dal** (Ia), ar-har dal

**aringa** (It), a-rin-ga

**aringa affumicata** (It), a-rin-ga af-fu-mi-ca-ta

**aringa marinata** (It), a-rin-ga ma-ri-na-ta

**arista** (It), a-ris-ta

**armadillo** (US), ar-ma-dil-lo

**armagnac** (Fr), ar-magn-ac

**Arme Ritter** (Gr), Arme Ritter

**armoricaine** (Fr), ar-mor-i-caine

**armut** (Tr), ar-mut

**arnab** (Ar), ar-nab

**arnabeet** (Ar), ar-na-beet

**arnavut ciğeri** (Tr), ar-na-vut ci-ğer-i

**arni** (Gk), ar-ni

**arni exohhiko** (Gk), ar-ni ex-o-khi-ko

**arni steen stamna** (Gk), ar-ni steen stam-na

**aromatic** (US), ar-o-mat-ic

**arraia** (Pg), a-rra-i-a

**arrayán** (Sp), a-rra-yán

**arreganato** (It), ar-re-ga-na-to

**arrôs** (Pg) (Brazil), a-rrôs

**arroser** (Fr), ar-ros-er

**arrostire sulla** (It), ar-ro-sti-re sul-la

**arrosto** (It), ar-ro-sto

**arrosto di agnello con patatine** (It), ar-ro-sto di a-gnel-lo con pa-ta-ti-ne

**arrowroot** (US), ar-row-root

**arroz** (Pg, Sp), a-rroz

*arroz abanda* (Sp), a-rroz a-ban-da

*arroz blanco con mejillones* (Sp), a-rroz blan-co con me-ji-llo-nes

*arroz con azafrán* (Sp), a-rroz con a-za-frán

*arroz con costra* (Sp), a-rroz con cos-tra

*arroz con frijoles* (Sp), a-rroz con fri-jo-les

*arroz con leche* (Sp), a-rroz con le-che

*arroz con pollo* (Mx), a-rroz con po-llo

*arroz doce* (Pg), a-rroz do-ce

*arroz refogado* (Pg), a-rroz re-fo-ga-do

*Art* (Gr), Art

*ärter* (Sw), är-ter

*ärter med fläsk* (Sw), är-ter med fläsk

*artichauts à la grecque* (Fr), ar-ti-chauts à la grecque

*artichauts à la vinaigrette* (Fr), ar-ti-chauts à la vin-ai-grette

*artichoke* (US), ar-ti-choke

*articsóka* (Hu), ar-ti-csó-ka

*Artischocken* (Gr), Ar-ti-schoc-ken

*Artischockenherzen* (Gr), Ar-ti-schoc-ken-her-zen

*artisjokk* (Nw), ar-ti-sjokk

*artisokka* (Fi), ar-ti-sok-ka

*ärtsoppa* (Sw), ärt-sop-pa

*artyčoky* (Cz), ar-ty-čo-ky

*aru* (Ia), a-ru

*arugula* (It), a-ru-gu-la

*arvi ki kari* (Ia), ar-vi ki ka-ri

*asadero* (Mx), a-sa-der-o

*asado* (Sp), a-sa-do

*asado de cerdo* (Sp), a-sa-do de cer-do

*asadura* (Sp), a-sa-du-ra

*asafetida* (Ia), a-sa-fe-ti-da

*asafetida poori* (Ia), a-sa-fe-ti-da poor-i

*asakusa nori* (Jp), a-sa-ku-sa no-ri

*asam* (In), a-sam

*asar* (Sp), a-sar

*asar a la lumbre* (Sp), a-sar a la lum-bre

*asar a la parrilla* (Sp), a-sar a la par-ri-lla

*asatsuki* (Jp), a-sat-su-ki

*ascalonia* (Sp), as-ca-lo-nia

*asciutta* (It), as-ciut-ta

*ascorbic acid* (US), a-scorb-ic ac-id

*aseer* (Ar), a-seer

*aseer burtuaan* (Ar), a-seer bur-tu-aan

*aseer il limoon* (Ar), a-seer il li-moon

*ash sak* (Ar), ash sak

*ashtarak tolma* (Rs), ash-tar-ak tol-ma

*Asiago* (It), A-si-a-go

*Asian pear* (US), A-sian pear

*asopao* (Sp-Puerto Rico), a-so-pao

*asparagi* (It), a-spa-ra-gi

*asparagi alla fiorentina* (it), a-spa-ra-gi al-la fio-ren-ti-na

*asparagus* (US), as-par-a-gus

*asparagus bean* (US), as-par-a-gus bean

*asparakopitas* (Gk), a-spar-a-ko-pi-tas

*asparges* (Da), a-sparges

*asparges* (Nw), a-spar-ges

*aspartame* (US), as-par-tame

*asperges* (Fr), as-per-ges

*asperges à la crème* (Fr), as-per-ges à la crème

*asperges en branches* (Fr), as-per-ges en bran-ches

*asperges mornay* (Fr), as-per-ges mor-nay

*asperges mousseline* (Fr), as-per-ges mousse-line

*aspergesoep* (Du), as-per-ge-soep

*asperillo* (Sp), as-pe-ri-llo

*aspic* (US), as-pic

*aspic jelly* (US), as-pic jel-ly

*aspro* (It), a-spro

*aspro krasi* (Gk), as-pro kra-si

*assado* (Pg), a-ssa-do

*assaisonnement* (Fr), as-sai-sonne-ment

*assaisonnement aromatique* (Fr), as-sai-sonne-ment ar-o-ma-tique

**Assam** (Ia), As-sam

*ässät* (Fi), äs-sät

*assida bil bufriwa* (Ar), as-si-da bil bu-fri-wa

*assiette anglaise* (Fr), as-siette ang-laise

*assiette assortie* (Fr), as-siette as-sor-tie

*assiette de charcuterie* (Fr), as-siette de char-cu-te-rie

*assortimento pazzo* (It), as-sor-ti-men-to paz-zo

*assortito* (It), as-sor-ti-to

*astaco* (It, Sp), a-sta-co

*astakos mayioneza* (Gk), a-sta-kos ma-yio-ne-za

*asuparagasu* (Jp), asu-par-a-gasu

*aşure* (Tr), a-şu-re

*ásványvizet* (Hu), ás-vány-vi-zet

*atalvina* (Sp), a-tal-vi-na

*atayef* (Ar), a-ta-yef

*atemoya* (US), at-e-moy-a

*athol brose* (Sc), ath-ol brose

*atjar* (In), at-jar

*atjar rebung* (In), at-jar re-bung

*atole* (Sp), a-to-le

*atpokat* (In), at-po-kat

*attereau* (Fr), at-ter-eau

*ättiksgurka* (Sw), åt-tiks-gur-ka

*atum* (Pg), a-tum

*atún* (Sp), a-tún

**Aubergine** (Gr), Au-ber-gi-ne

*aubergine* (GB, Fr, It), au-ber-gine

*aubergines à la niçoise* (Fr), au-ber-gines à la ni-çoise

*au beurre* (Fr), au beurre

*au bleu* (Fr), au bleu

**Auflauf** (Gr), Auf-lauf

*au four* (Fr), au four

**Aufschnitt** (Gr), Auf-schnitt

*au gratin* (Fr), au gra-tin

*augurken* (Du), au-gur-ken

*au jus* (Fr), au jus

*au lait* (Fr), au lait

*au lard* (Fr), au lard

*au naturel* (Fr), au na-tur-el

*aure* (Nw), au-re

*aurore, sauce* (Fr), au-rore, sauce

*au sang* (Fr), au sang

*Austern* (Gr), Aus-tern

*auszpik* (Po), ausz-pik

*avakkai mangai* (Ia), a-vak-kai man-gai

*avêia* (Pg), a-vê-i-a

*aveline* (Fr), av-e-line

*avellana* (It, Sp), a-ve-lla-na

*aves* (Sp), av-es

*avgha* (Gk), av-gha

*avgolemono* (Gk), av-go-le-mo-no

*avkokt torsk* (Nw), av-kokt torsk

*avocado* (US), av-o-ca-do

*avocat* (Fr), av-o-cat

*avocat farci de crevettes* (Fr), av-o-cat far-ci de cre-vettes

*awabi* (Jp), a-wa-bi

*ayam* (In), a-yam

*ayam panggang bumbu besengek* (In), a-yam pang-gang bum-bu be-se-ngek

*ayam panike* (In), a-yam pan-i-ke

*ayam percik* (Ml), a-yam per-cik

*ayam tauco* (In), a-yam tau-co

*ayran* (Tr), ay-ran

*ayskrimu* (Af-Swahili), ays-kri-mu

*ayskrimu ya vanila* (Af-Swahili), ays-kri-mu ya va-ni-la

*ayu* (Jp), a-yu

*az* (Tr), az

*azafrán* (Sp), a-za-frán

*azarole* (US), az-a-role

*azedinha* (Pg), a-ze-di-nha

*azedo* (Pg), a-ze-do

*Azeitão* (Pg), A-ze-i-tã-o

*azeite* (Pg), a-ze-i-te

*azeitonas* (Pg), a-ze-i-to-nas

*azijn* (Du), a-zijn

*azu* (Rs), a-zu

*azúcar* (Sp), a-zú-car

*azuki* (Jp), a-zu-ki

*azukian* (Jp), a-zuk-i-an

*azyme* (Fr), a-zyme

# B

*baars* (Du), baars

*bab* (Hu), bab

*baba* (Fr), ba-ba

*baba au rhum* (Fr), ba-ba au rhum

*babaco* (Sp), ba-ba-co

*baba ghannoug* (Ar), ba-ba
   ghan-noug
*babassu* (Pg), ba-bas-su
*babat* (In), ba-bat
*bā bău dŭng-gwā tāng* (Ch),
   bā bău dūng-gwā tāng
*babeczki smietankowe* (Po),
   ba-becz-ki smie-tan-kowe
*babeurre* (Fr), ba-beurre
*babgulyas* (Hu), bab-gu-lyas
*babi* (In), ba-bi
*babi asam pedas* (In), ba-bi
   a-sam pe-das
*babka* (Po), bab-ka
*bableves* (Hu), bab-le-ves
*bableves csipetkével* (Hu),
   bab-le-ves csi-pet-ké-vel
*bábovka* (Cz), bá-bov-ka
*bacalao* (Sp), ba-ca-lao
*bacalao al ajo arriero* (Sp),
   ba-ca-lao al a-jo ar-ri-e-ro
*bacalao al pil-pil* (Sp), ba-
   ca-lao al pil-pil
*bacalhau* (Pg), ba-ca-lha-u
*baccalà* (It), bac-ca-là
*back bacon* (GB), back bacon
*Backhendl* (Gr-Austria),
   Back-hen-dl
*Backobst* (Gr), Back-obst
*Backpflaumen* (Gr), Back-
   pflau-men
*Backwerk* (Gr), Back-werk
*bacon* (US), ba-con
*bacon* (Nw), ba-con
*bacon* (Fr), ba-con
*bacon med ägg* (Sw), ba-con
   med ägg
*bacon og æg* (Da), ba-con og
   æg

*bacon rasher* (GB), ba-con
   rash-er
*bacon strip* (US), ba-con strip
*badakelu vinjal* (Ia), ba-da-
   ke-lu vin-jal
*badam* (Ia), ba-dam
*badam barfi* (Ia), ba-dam
   bar-fi
*badami pasanda* (Ia), ba-
   dam-i pa-san-da
*badderlocks* (US), bad-der-
   locks
*bær* (Nw), bær
*bagel* (Jw), ba-gel
*bagel chip* (US), ba-gel chip
*baghari jhinga* (Ia), ba-gha-
   ri jhin-ga
*bagna cauda* (It), bag-na
   cau-da
*bagnare* (It), bag-na-re
*Bagnes* (Fr-Swiss), Ba-gnes
*bagoong* (Ph), ba-goong
*bagt kartoffel* (Da), bagt kar-
   toff-el
*baguette* (Fr), ba-guette
*bái-cài* (Ch), bái-cài
*baicoli* (It), ba-i-co-li
*baidakov kulebiaka* (Rs),
   bai-da-kov ku-le-bia-ka
*baies* (Fr), baies
*bái-gwo* (Ch), bái-gwo
*bái-lwó-bwō* (Ch), bái-lwó-
   bwō
*baingan* (Ia), bain-gan
*baingan bharta* (Ia), baingan bhar-ta
*baingan pakora* (Ia), baingan pa-ko-ra
*bain-marie* (Fr), bain-ma-rie
*baiser* (Fr), bai-ser

**bái-shŭ** (Ch), bái-shŭ
**bái-tsài** (Ch), bái-tsài
**bái-tsài bai yú-ywán** (Ch), bái-tsài bai yú-ywán
**bái-yú tāng** (Ch), bái-yú tāng
**bajia** (Af-Swahili), ba-ji-a
**bajra** (Ia), baj-ra
**bakalar** (SC), ba-ka-lar
**bakaliaros** (Gk), ba-ka-lia-ros
**bakarkhani** (Ia), ba-kar-kha-ni
**bake, to** (US), bake, to
**bakeapple** (Ca), bake-ap-ple
**bake blind, to** (GB), bake blind, to
**baked Alaska** (US), baked A-las-ka
**bakelse tart** (Sw), bak-el-se tart
**baker's cheese** (US), bak-er's cheese
**baking powder** (US), bak-ing pow-der
**baking soda** (US), bak-ing so-da
**baklava** (Ar, Gk, Tr), bak-la-va
**baklazhan** (Rs), ba-kla-zhan
**baklazhan s ovoshami** (Rs), ba-kla-zhan s o-vo-sha-mi
**bak mie** (In), bak mie
**bak pao** (In), bak pao
**bakré ka gosht** (Ia), bak-ré ka gosht
**balachan** (Ml), ba-la-chan
**balachong** (Ia), ba-la-chong
**balah** (Ar), ba-lah
**balik izgara** (Tr), ba-lik iz-ga-ra

**balik tavasi** (Tr), ba-lik ta-va-si
**ballon** (Fr), bal-lon
**ballottine** (Fr), bal-lot-tine
**baloney** (US), ba-lo-ney
**balsamella** (It), bal-sa-mel-la
**balsamic vinegar** (US), bal-sam-ic vin-e-gar
**Balsamkraut** (Gr), Bal-sam-kraut
**balsam pear** (US), bal-sam pear
**Baltic herring** (US), Bal-tic her-ring
**balungi** (Af-Swahili), ba-lu-ngi
**balushahi** (Ia), ba-lu-sha-hi
**balyk i siomga** (Rs), ba-lyk i siom-ga
**bamboo shoots** (US), bam-boo shoots
**bamia** (Af-Swahili), ba-mi-a
**bamja** (SC), bam-ja
**bamya** (Ar), bam-ya
**ban** (Ia), ban
**banaani** (Fi), ba-naa-ni
**banan** (Nw, Rs, Sw), ba-nan
**banán** (Cz, Hu), ba-nán
**banana** (US), ba-na-na
**banana pepper** (US), ba-na-na pep-per
**bananas Foster** (US), ba-na-nas Fos-ter
**banana split** (US), ba-na-na split
**banane** (It, Gr), ba-na-ne
**banane** (Fr), ba-nane
**bananes à crème chantilly** (Fr), ba-nanes à crème chan-til-ly

*bananes flambées* (Fr), ba-nanes flam-bées

*Banbury cake* (GB), Ban-bur-y cake

*bancha* (Jp), coarse green tea

*band gobbi* (Ia), band go-bhi

*banger* (GB), bang-er

*bangers and mash* (GB), bang-ers and mash

*banh cuon* (Vt), banh cu-on

*banh xeo* (Vt), banh xeo

*banitsa* (Bu), ba-ni-tsa

*bankebiff* (Nw), ban-ke-biff

*bankekød* (Da), ban-ke-kød

*bàn ming-bá* (Ch), bàn ming-há

*bannock* (Sc), ban-nock

*Banon* (Fr), Ba-non

*bàn-shù de jī-dàn* (Ch), bàn-shù de jī-dàn

*bap* (Sc), bap

*bar* (Fr), bar

*bär* (Sw), bär

*barackleves* (Hu), ba-rack-le-ves

*barackpálinkát* (Hu), ba-rack-pá-lin-kát

*baranii bok s kashei* (Rs), ba-ra-nii bok s ka-shei

*baranina* (Po), ba-ra-ni-na

*bárány* (Hu), bá-rány

*báránypörkölt* (Hu), bá-rány-pör-költ

*barashek* (Rs), ba-ra-shek

*barbabietole* (It), bar-ba-bie-to-le

*barbacoa* (Mx), bar-ba-co-a

*Barbados cherry* (Cb), Bar-ba-dos cher-ry

*Barbados sugar* (Cb), Bar-ba-dos sug-ar

*barbecue* (US), bar-be-cue

*barbecue sauce* (US), bar-be-cue sauce

*barbel* (US), bar-bel

*barberry* (US), bar-ber-ry

*barbes-de-capuchin* (Fr), barbes-de-cap-u-chin

*barbue* (Fr), bar-bue

*barbunia* (Gk), bar-bu-nia

*bardana* (It), bar-da-na

*bardé* (Fr), bar-dé

*barfi* (Ia), bar-fi

*bar grillé* (Fr), bar gri-llé

*barigoule* (Fr), ba-ri-goule

*barkoukess* (Ar), bar-kou-kess

*bar-le-Düc* (Fr), bar-le-Düc

*barley* (US), bar-ley

*barley flour* (US), bar-ley flour

*barley sugar* (GB), bar-ley sug-ar

*barmbrack* (Ir), barm-brack

*Bärme* (Gr), Bär-me

*barnacle* (US), bar-na-cle

*barna kenyér* (Hu), bar-na ken-yér

*baron* (US), bar-on

*baroo* (Ar), bar-oo

*bar pochè á l'oiseille* (Fr), bar pochè á l'oi-seille

*barquette* (Fr), bar-quette

*barquettes ostendaise* (Fr), bar-quettes os-ten-daise

*barracuda* (US), bar-ra-cu-da

*bar raye* (Fr), bar raye

*Barsch* (Gr), Barsch

*barszez* (Po), bar-szez

*basal* (Ar), ba-sal
*basar* (Jw), ba-sar
*basar bakar* (Jw-Israel), ba-sar ba-kar
*basar egel* (Jw-Israel), ba-sar e-gel
*basar keves* (Jw-Israel), ba-sar ke-ves
*basil* (US), bas-il
*basilic* (Fr), ba-si-lic
*basilico* (It), ba-si-li-co
*Basilienkraut* (Gr), Ba-sil-ien-kraut
*basmati rice* (Ia), bas-ma-ti rice
*basquaise* (Fr), bas-quaise
*bass* (US), bass
*baste, to* (US), baste, to
*batā* (Jp), ba-tā
*bata* (Af-Swahili), ba-ta
*bâtarde, sauce* (Fr), bâ-tarde, sauce
*batata* (Ar), ba-ta-ta
*batatas* (Pg), ba-ta-tas
*batatas* (Sp), ba-ta-tas
*batatas doces* (Pg), ba-ta-tas do-ces
*batatis* (Ar), ba-ta-tis
*batatis maleeya* (Ar), ba-ta-tis ma-lee-ya
*bata wa bukini* (Af-Swahili), ba-ta wa bu-ki-ni
*batér* (Ia), ba-tér
*Bath bun* (GB), Bath bun
*batido* (Sp), ba-ti-do
*bâtonnets* (Fr), bâ-ton-nets
*battak* (Ia), bat-tak
*batteekh* (Ar), bat-teekh
*batter* (US), bat-ter

*batter bread* (US), bat-ter bread
*battre* (Fr), bat-tre
*baudroie* (Fr), bau-droie
*bāu-dz* (Ch), bāu-dz
*Bauernart* (Gr), Bau-ern-art
*Bauernbrot* (Gr), Bau-ern-brot
*Bauernsuppe* (Gr), Bau-ern-sup-pe
*bauletto* (It), bau-let-to
*Baumtorte* (Gr), Baum-tor-te
*Baumwollöl* (Gr), Baum-woll-öl
*bāu-yú* (Ch), bāu-yú
*Bavarian cream* (US), Ba-var-i-an cream
*bavarois* (Fr), ba-va-rois
*bawang* (In), ba-wang
*bawang putih* (In), ba-wang pu-tih
*bawd* (Sc), bawd
*bayam* (In), ba-yam
*bayd* (Ar), bayd
*bayd masloo* (Ar), bayd mas-loo
*bay leaf* (US), bay leaf
*bayonnaise, à la* (Fr), ba-yon-naise, à la
*bay scallop* (US), bay scal-lop
*bažant* (Cz), ba-žant
*bean* (US), bean
*bean curd* (US), bean curd
*bean sprouts* (US), bean sprouts
*bean thread noodles* (US), bean thread noo-dles
*beard* (US), beard
*béarnaise, sauce* (Fr), bé-ar-naise, sauce

**beat, to** (US), beat, to
**Beaumont** (Fr), Beau-mont
**becada** (Sp), be-ca-da
**bécasse** (Fr), bé-casse
**bécassine** (Fr), bé-cas-sine
**beccaccia** (It), bec-cac-cia
**beccaccino** (It), bec-cac-ci-no
**beccaccino allo spiedo** (It), bec-cac-ci-no al-lo spi-e-do
**béchamel, sauce** (Fr), bé-cha-mel, sauce
**bêche de mer** (Fr), bêche de mer
**beckasin** (Sw), bec-ka-sin
**bécsi szelet** (Hu), bé-csi sze-let
**beebeek** (Ia), bee-beek
**beechnut** (US), beech-nut
**beef** (US), beef
**beef à la mode** (US), beef à la mode
**beef Stroganoff** (US), beef Stro-ga-noff
**beef tartar** (US), beef tar-tar
**beef tea** (US), beef tea
**beef Wellington** (US), beef Wel-ling-ton
**beer** (US), beer
**beer cheese** (US), beer cheese
**beet** (US), beet
**beetroot** (GB), beet-root
**beet sugar** (US), beet sug-ar
**befsztyk** (Po), bef-sztyk
**befsztyk tartarski** (Po), bef-sztyk tar-tar-ski
**beignets** (Fr), bei-gnets
**beignets de pommes** (Fr), bei-gnets de pommes

**beignets niçoise** (Fr), bei-gnets ni-çoise
**Béi-jīng fén** (Ch), Béi-jīng fén
**Béi-jīng jyău-dz** (Ch), Béi-jīng jyău-dz
**beinasu iradashi** (Jp), be-i-na-su i-ra-da-shi
**Bekassine** (Gr), Be-kas-si-ne
**bēkon to tamago** (Jp), bacon and eggs.
**belegde broodjes** (Du), be-leg-de brood-jes
**belegtes Brot** (Gr), be-leg-tes Brot
**Belgian endive** (US), Belgian en-dive
**beli luk** (SC), be-li luk
**belimbing** (In), belim-bing
**belle-dijonnaise, à la** (Fr), belle-di-jon-naise, à la
**bellevue** (Fr), belle-vue
**bell pepper** (US), bell pep-per
**belon** (Fr), be-lon
**belo vino** (SC), be-lo vi-no
**Bel Paese** (It), Bel Pa-e-se
**beluga caviar** (US), be-lu-ga cav-i-ar
**ben cotto** (It), ben cot-to
**benishoga** (Jp), be-ni-sho-ga
**benløse fugle** (Da), ben-lø-se fug-le
**benne seeds** (US), benne seeds
**berberé** (Af), ber-ber-é
**Bercy, à la** (Fr), Ber-cy, à la
**berenjena** (Sp), be-ren-je-na
**bergamot** (US), ber-ga-mot
**bergère** (Fr), ber-gère
**Bergkäse** (Gr), Berg-kä-se

**beringela** (Pg-Brazil), be-rin-ge-la

**Berliner Pfannkuchen** (Gr), Ber-lin-er Pfann-ku-chen

**Bermuda onion** (US), Ber-mu-da on-ion

**Berner Sauce** (Gr), Bern-er Sau-ce

**berro** (Sp), ber-ro

**berry** (US), ber-ry

**berry sugar** (US), ber-ry sug-ar

**berza rizada** (Sp), ber-za ri-za-da

**besan** (Ia), be-san

**beschuit** (Du), be-schuit

**beschuittaart** (Du), be-schuit-taart

**besciamella** (It), be-scia-mel-la

**besengek daging sapi** (In), be-se-ngek da-ging sa-pi

**bessenvla** (Du), bes-sen-vla

**besugo al horno** (Sp), be-su-go al hor-no

**beta-carotene** (US), be-ta-car-o-tene

**betelloh** (Ar), be-tel-loh

**betteraves** (Fr), bet-te-raves

**betteraves à l'orange** (Fr), bet-te-raves à l'or-ange

**beurre** (Fr), beurre.

**beurre à la maitre d'hôtel** (Fr), beurre à la mai-tre d'hô-tel

**beurre blanc** (Fr), beurre blanc

**beurre d'ail** (Fr), beurre d'ail

**beurre d'anchois** (Fr), beurre d'an-chois

**beurre d'échalote** (Fr), beurre d'é-cha-lote

**beurre de citron** (Fr), beurre de ci-tron

**beurre de crevettes** (Fr), beurre de cre-vettes

**beurre fondu** (Fr), beurre fon-du

**beurre manié** (Fr), beurre man-ié

**beurre noir** (Fr), beurre noir

**beurre vert** (Fr), beurre vert

**beyaz peynir** (Tr), bey-az pey-nir

**beyin** (Tr), bey-in

**bezelye** (Tr), be-zel-ye

**bharta** (Ia), bhar-ta

**bhéja** (Ia), bhé-ja

**bhelpuri** (Ia), bhel-pu-ri

**bhindi** (Ia), bhin-di

**bhujiya** (Ia), bhu-ji-ya

**bhuna chaval** (Ia), bhu-na cha-val

**bhuna hua** (Ia), bhu-na hu-a

**bialy** (Po), bi-a-ly

**bianchette di vitello** (It), bian-chette di vi-tel-lo

**bianchetti** (It), bian-chet-ti

**bianco** (It), bian-co

**bianco de Spagna** (It), bian-co de Spag-na

**bianco d'uovo** (It), bian-co d'uo-vo

**Bibb lettuce** (US), Bibb let-tuce

**biber** (SC, Tr), bi-ber

**biber dolmasi** (Tr), bi-ber dol-ma-si

**bibimpap** (Kr), bi-bim-pap

**bibita** (It), bi-bi-ta

**bicarbonate of soda** (US), bi-car-bo-nate of soda
**bidingehn** (Ar), bi-din-gehn
**bief** (Du), bief
**biefstuk** (Du), bief-stuk
**bien asado** (Sp), bien a-sa-do
**bien cuit** (Fr), bien cuit
**Bienenstich** (Gr), Bie-nen-stich
**bier** (Du), bier
**Bier** (Gr), Bier
**bière** (Fr), bière
**Bierkaltschale** (Gr), Bier-kalt-scha-le
**Bierkäse** (Gr), Bier-kä-se
**Bierwurst** (Gr), Bier-wurst
**bieten met appelen** (Du), biet-en met ap-pel-en
**bietola da coste** (It), bi-e-to-la da co-ste
**bife** (Pg), bi-fe
**biff** (Nw), biff
**biff à la Lindström** (Sw), biff à la Lind-ström
**biffstek** (Sw), biff-stek
**bifsztek** (Hu), bif-sztek
**biftec** (Sp), bif-tec
**biftec de ternera** (Sp), bif-tec de ter-ne-ra
**bifteck** (Fr), bif-teck
**bifteck entrecôte** (Fr), bif-teck en-tre-côte
**bifteck pavé au poivre** (Fr), bif-teck pa-vé au poivre
**biftek** (Cz, Tr), bif-tek
**bifuteki** (Jp), bi-fu-te-ki
**bigarade, sauce** (Fr), bi-ga-rade, sauce
**bigaro** (Sp), bi-ga-ro

**bignè** (It), bi-gnè
**bignè al formaggio** (It), bi-gnè al for-mag-gio
**bigorneau** (Fr), bi-gor-neau
**bigos** (Po), bi-gos
**bibon** (Ph), bi-hon
**biksemad** (Da), bik-se-mad
**bilberry** (GB), bil-ber-ry
**bílé vino** (Cz), bí-lé vi-no
**billi-bi** (Fr), bil-li-bi
**biltong** (Af), bil-tong
**binatang buruan** (In), bi-na-tang bu-ru-an
**binder** (US), bind-er
**bing** (Ch), bing
**Bing cherry** (US), Bing cher-ry
**bing-gān** (Ch), bing-gān
**bīng-gwo-dz-lù** (Ch), bīng-gwo-dz-lù
**bīng-jī-líng** (Ch), bīng-jī-líng
**biotin** (US), bi-o-tin
**bira** (Bu), bi-ra
**bird pepper** (US), bird pep-per
**bird's nest** (US), bird's nest
**bird's nest soup** (US), bird's nest soup
**biringani** (Af-Swahili), bi-ri-nga-ni
**Birne** (Gr), Bir-ne
**birra** (It), bir-ra
**biryani** (Ia), bir-ya-ni
**biscoitos** (Pg), bis-coi-tos
**biscotte** (Fr), bis-cotte
**biscotti** (It), bi-scot-ti
**biscotti all' anice** (It), bi-scot-ti al-l' a-ni-ce
**biscottini di mandorle** (It), bi-scot-ti-ni di man-dor-le

*biscotto tortoni* (It), bi-scot-to tor-to-ni
*biscuit* (US), bis-cuit
*biscuit* (GB), bis-cuit
*biscuit* (Fr), bis-cuit
*bisilla* (Ar), bi-sil-la
*biskopskake* (Nw), bis-kops-ka-ke
*Biskuit* (Gr), Bis-kuit
*biskuti* (Af-Swahili), bi-sku-ti
*Bismarck herring* (Gr), Bismarck her-ring
*bisque* (Fr), bisque
*bisque d'ecrevisse* (Fr), bisque d'e-cre-visse
*bistecca* (It), bi-stec-ca
*bistecca alla Fiorentina* (It), bi-stec-ca al-la Fio-ren-ti-na
*bistik* (In), bi-stik
*bisuketto* (Jp), bi-su-ket-to
*bitamin* (Jp), bi-ta-min
*bitok* (Rs), bi-tok
*bitter almond* (US), bit-ter al-mond
*bitterballen* (Du), bit-ter-ball-en
*bitter chocolate* (US), bit-ter choc-o-late
*bitterkoekjes* (Du), bit-ter-koek-jes
*bitterkoekjesvla* (Du), bit-ter-koek-jes-vla
*Bittermandel* (Gr), Bit-ter-man-del
*bitter melon* (US), bit-ter mel-on
*bitter orange* (US), bit-ter or-ange

*bitters* (GB), bit-ters
*biwa* (Jp), bi-wa
*bixin* (US), bix-in
*bizcochada* (Sp), biz-co-cha-da
*bizcocho* (Sp), biz-co-cho
*bizcocho genovesa* (Sp), biz-co-cho ge-no-ve-sa
*björnbär* (Sw), björn-bär
*bjørnebær* (Nw), bjør-ne-bær
*blåbær* (Nw), blå-bær
*blåbærpannekake* (Nw), blå-bær-pan-ne-ka-ke
*blåbærsuppe* (Da), blå-bær-sup-pe
*blåbär* (Sw), blå-bär
*black bean* (US), black bean
*blackberry* (US), black-ber-ry
*black-bottom pie* (US), black-bot-tom pie
*blackened* (US), black-ened
*black-eyed peas* (US), black-eyed peas
*black pepper* (US), black pep-per
*black pudding* (GB), black pud-ding
*black salsify* (US), black sal-si-fy
*black sea bass* (US), black sea bass
*blackstrap molasses* (US), black-strap mo-las-ses
*black treacle* (GB), black trea-cle
*black walnut* (US), black wal-nut
*bladselleri* (Da), blad-sel-le-ri
*blaeberry* (Sc), blae-ber-ry

*blanc de cuisson* (Fr), blanc de cuis-son

*blanch* (US), blanch

*blancmange* (Fr), blanc-mange

*blandad* (Sw), blan-dad

*blanquette* (Fr), blan-quette

*blanquette d'agneau à l'an-cienne* (Fr), blan-quette d'agn-eau à l'an-cienne

*blanquette de veau* (Fr), blan-quette de veau

*bláthach* (Ir), bláth-ach

*Blaubeere* (Gr), Blau-bee-re

*blaze* (US), blaze

*blé* (Fr), blé

*bleak* (US), bleak

*blend* (US), blend

*blé noir* (Fr), blé noir

*bleu* (Fr), bleu

*blewah* (In), ble-wah

*blinde vinken* (Du), blin-de vin-ken

*blind Huhn* (Gr), blind Huhn

*blini* (Rs), bli-ni

*blintze* (Jw), blint-ze

*bloater* (GB), bloat-er

*blodbudding* (Da), blod-bud-ding

*blodfersk* (Nw), blod-fersk

*blødkogte aeg* (Da), blød-kog-te aeg

*blodkorv* (Sw), blod-korv

*blodpølse* (Da), blod-pøl-se

*bloedworst* (Du), bloed-worst

*bloemkool* (Du), bloem-kool

*blomkaal* (Da), blom-kaal

*blomkål* (Da, Nw, Sw), blom-kål

*blommer* (Da), blom-mer

*blond de veau* (Fr), blond de veau

*blond de volaille* (Fr), blond de vo-laille

*blondir* (Fr), blon-dir

*blood* (US), blood

*blood orange* (US), blood or-ange

*blood sausage* (US), blood sau-sage

*bløtkake* (Nw), bløt-ka-ke

*bløtkokt* (Nw), bløt-kokt

*blueberry* (US), blue-ber-ry

*blue cheese* (US), blue cheese

*blue cheese dressing* (US), blue cheese dress-ing

*blue crab* (US), blue crab

*bluefish* (US), blue-fish

*Blumenkohl* (Gr), Blu-men-kohl

*Blutwurst* (Gr), Blut-wurst

*bobi* (Rs), bo-bi

*bobotee* (US), bo-bo-tee

*bobotie* (Af), bo-bo-tie

*böbrek* (Tr), bö-brek

*bobwhite* (US), bob-white

*bocadillo* (Sp), bo-ca-dillo

*bocconcini* (It), boc-con-ci-ni

*Bock* (Gr), Bock

*böckling* (Sw), böck-ling

*Bocksbart* (Gr), Bocks-bart

*Bockwurst* (Gr), Bock-wurst

*bodega* (Sp), bo-de-ga

*boerenkool* (Du), boer-en-kool

*boerenkool met worst* (Du), boer-en-kool met worst

*boeuf* (Fr), boeuf

**boeuf à la mode** (Fr), boeuf à la mode

**boeuf bouilli** (Fr), boeuf bou-illi

**boeuf bourguignon** (Fr), boeuf bour-gui-gnon

**boeuf en gelée** (Fr), boeuf en ge-lée

**boeuf miroton** (Fr), boeuf mi-ro-ton

**boeuf rôti** (Fr), boeuf rô-ti

**boeuf salé** (Fr), boeuf sa-lé

**bøf med løg** (Da), bøf med løg

**bøf tartar** (Da), bøf tar-tar

**Bohnen** (Gr), Boh-nen

**boil** (US), boil

**boiler** (US), boil-er

**bok choy** (Ch), bok choy

**bokking** (Du), bok-king

**bola** (Mx), bo-la

**bolacha** (Pg), bo-la-cha

**bolets** (Fr), bo-lets

**boletus** (US), bo-le-tus

**bolillo** (Mx), bo-li-llo

**bolinchos de bacalhau** (Pg), bo-lin-chos de ba-ca-lha-u

**bolle** (Da, Nw), bol-le

**boller** (Da), bol-ler

**bollito** (It), bol-li-to

**bollito misto** (It), bol-li-to mis-to

**bollo** (Sp), bo-llo

**bolltee** (Ar), boll-tee

**bôlo** (Pg), bô-lo

**bologna** (It), bo-lo-gna

**Bolognese, alla** (It), Bo-lo-gnese, al-la

**bombas de camarones** (Sp), bom-bas de ca-ma-ro-nes

**Bombay duck** (Ia), Bom-bay duck

**bombe glacée** (Fr), bombe gla-cée

**bombons** (Pg), bom-bons

**bommaloe** (Ia), bom-ma-loe

**böna** (Sw), bö-na

**bonbon** (Fr), bon-bon

**bondas** (Ia), bon-das

**bondpige med slør** (Da), bond-pi-ge med slør

**bone, roll and tie** (US), bone, roll and tie

**bonen** (Du), bo-nen

**bonensla** (Du), bo-nen-sla

**boniatos** (Sp), bo-nia-tos

**boniatos confitadas** (Sp), bo-nia-tos con-fi-ta-das

**boning** (US), bon-ing

**bonite à dos rayé** (Fr), bon-ite à dos rayé

**bonito** (US), bo-ni-to

**bonne-dame** (Fr), bonne-dame

**bonne femme, à la** (Fr), bonne femme, à la

**bønner** (Da, Nw), bøn-ner

**bönor** (Sw), bö-nor

**boonchi** (Cb), boon-chi

**boquerónes** (Sp), bo-que-ró-nes

**borage** (US), bor-age

**boranija** (SC), bo-ra-ni-ja

**borç** (Tr), borç

**bordelaise, à la** (Fr), bor-de-laise, à la

**börek** (Tr), bö-rek

**borjúhús** (Hu), bor-jú-hús

**borjúpörkölt** (Hu), bor-jú-pör-költ

**borjúvelö** (Hu), bor-jú-ve-lö
**borleves** (Hu), bor-le-ves
**borrego** (Pg), bo-rre-go
**bors** (Hu), bors
**borsch** (Rs), borsch
**borscht** (Po), borscht
**borsó** (Hu), bor-só
**borsóleves** (Hu), bor-só-le-ves
**borstplaat** (Du), borst-plaat
**borststuk** (Du), borst-stuk
**borůvky** (Cz), bo-rův-ky
**bosbessen** (Du), bos-bes-sen
**Boscaiola, alla** (It), Bos-ca-io-la, al-la
**Boston baked beans** (US), Bos-ton baked beans
**Boston brown bread** (US), Bos-ton brown bread
**Boston lettuce** (US), Bos-ton let-tuce
**boszorkánybab** (Hu), bo-szor-kány-hab
**bot** (Du), bot
**botargo** (It), bo-tar-go
**boter** (Du), bo-ter
**boterhamkoek** (Du), bo-ter-ham-koek
**botifarra amb mongetes** (Sp), bo-ti-far-ra amb mon-ge-tes
**boti kabab** (Ia), bo-ti ka-bab
**botvinya** (Rs), bot-vin-ya
**botwina** (Po), bot-wi-na
**boubliki** (Rs), bou-bli-ki
**bouchée** (Fr), bou-chée
**bouchée à la reine** (Fr), bou-chée à la reine
**bouchère, sauce** (Fr), bou-chère, sauce

**boudin** (Fr), bou-din
**boudin blanc** (Fr), bou-din blanc
**boudin noir** (Fr), bou-din noir
**bouillabaisse** (Fr), bouil-la-baisse
**bouilli** (Fr), bou-illi
**bouillon** (Fr), bou-illon
**bouillon de boeuf** (Fr), bou-illon de boeuf
**boulangère, à la** (Fr), bou-lan-gère, à la
**boule-de-neige** (Fr), boule-de-neige
**boulette** (Fr), bou-lette
**bouquet** (Fr), bou-quet
**bouquet garni** (Fr), bou-quet gar-ni
**bouquetière, à la** (Fr), bou-que-tière, à la
**bourgeoise, à la** (Fr), bourge-oise, à la
**bourguignon, à la** (Fr), bour-gui-gnon, à la
**bourride** (Fr), bour-ride
**boysenberry** (US), boy-sen-ber-ry
**Brachsenmakrele** (Gr), Brach-sen-mak-re-le
**braciola** (It), bra-cio-la
**braciola di maiale** (It), bra-cio-la di ma-ia-le
**braciola di manzo** (It), bra-cio-la di man-zo
**braciolette d'abbacchio** (It), bra-cio-lette d'ab-bac-chio
**bräckkorv** (Sw), bräck-korv
**bradán** (Ir), bra-dán
**braga** (Rs), bra-ga

*brain* (US), brain
*braise, to* (US), braise, to
*braisé* (Fr), brai-sé
*braisé de boeuf* (Fr), brai-sé de boeuf
*bramble* (US), bram-ble
*brambor* (Cz), bram-bor
*bramborák* (Cz), bram-bo-rák
*bramborová kaše* (Cz), bram-bo-ro-vá ka-še
*bramborová polévka* (Po), bram-bo-ro-vá po-lév-ka
*bramborový knedlík* (Cz), bram-bo-ro-vý kned-lík
*bramborový salát* (Cz), bram-bo-ro-vý sa-lát
*bran* (US), bran
*brancin* (SC), bran-cin
*brandade* (Fr), bran-dade
*brandewijn* (Du), bran-de-wijn
*brandy* (US), bran-dy
*branzino* (It), bran-zi-no
*brasato* (It), bra-sa-to
*braskartofler* (Da), bras-kar-tof-ler
*brasserie* (Fr), bras-ser-ie
*Braten* (Gr), Bra-ten
*Bratfisch* (Gr), Brat-fisch
*Brathänchen* (Gr), Brat-hän-chen
*Bratheringe* (Gr), Brat-he-ring-e
*Bratkartoffeln* (Gr), Brat-kar-tof-feln
*Bratwurst* (Gr), Brat-wurst
*Bräune Tunke* (Gr), Bräune Tun-ke

*Braunschweiger* (Gr), Braun-schwei-ger
*brawn* (Gr), brawn
*Brazil nut* (US), Bra-zil nut
*brazo mercedes* (Ph), bra-zo mer-ce-des
*breac geal* (Ir), breac geal
*bread* (US), bread
*bread, to* (US), bread, to
*bread and butter pickles* (US), bread and but-ter pick-les
*breadfruit* (US), bread-fruit
*bread pudding* (US), bread pud-ding
*bread sauce* (GB), bread sauce
*bread sticks* (US), bread sticks
*bream* (US), bream
*brécol* (Sp), bré-col
*bredflab* (Da, Nw), bred-flab
*bree* (Sc), bree
*Brei* (Gr), Brei
*brème de mer* (Fr), brème de mer
*brennesnut* (Nw), bren-nes-nut
*breskva* (SC), bre-skva
*bretonne, à la* (Fr), bre-tonne, à la
*brew* (US), brew
*brewat* (Ar), bre-wat
*brewer's yeast* (US), brew-er's yeast
*Brezel* (Gr), Bre-zel
*brick cheese* (US), brick cheese
*brick tea* (US), brick tea
*Brie* (Fr), Brie

*brigidini* (It), bri-gi-di-ni
*brik* (Ar), brik
*brik bil lahm* (Ar), brik bil lahm
*brill* (US), brill
*brine* (US), brine
*bringa* (Sw), brin-ga
*bringebær* (Nw), brin-ge-bær
*brinjal* (Ia), brin-jal
*brioche* (Fr), bri-oche
*brioche de fois gras* (Fr), brioche de fois gras
*brioška* (Cz), bri-oš-ka
*brisket* (US), bris-ket
*brisler* (Da), bris-ler
*brisling* (Da, Nw), bris-ling
*broa* (Pg), bro-a
*broad bean* (US), broad bean
*broccoletti di Brusselle* (It), broc-co-let-ti di Brus-sel-le
*broccoli* (US), broc-co-li
*broccoli* (It), broc-co-li.
*broccoli al formaggio* (It), broc-co-li al for-mag-gio
*broccoli all' agro* (It), broc-co-li al-l' a-gro
*brochan* (Sc), broch-an
*broche, à la* (Fr), broche, à la
*brochet* (Fr), bro-chet
*brochette* (Fr), bro-chette
*brochette de mouton* (Fr), bro-chette de mou-ton
*bröd* (Sw), bröd
*brød* (Da, Nw), brød
*brodetto* (It), bro-det-to
*brodetto di pesce alla veneziana* (It), bro-det-to di pes-ce al-la ve-ne-zi-a-na
*brodo* (It), bro-do

*brodo di manzo* (It), bro-do di man-zo
*brodo di pollo* (It), bro-do di pol-lo
*brødsuppe* (Da), brød-sup-pe
*broil* (US), broil
*broiler* (US), broil-er
*brood* (Du), brood
*broodjes* (Du), brood-jes
*broodpap* (Du), brood-pap
*broqueta* (Sp), bro-que-ta
*Bröschen* (Gr), Brö-schen
*brose* (Sc), brose
*broskev* (Cz), bros-kev
*Brot* (Gr), Brot
*Brotaufstrich* (Gr), Brot-auf-strich
*Brötchen* (Gr), Bröt-chen
*broth* (US), broth
*Brotkoch* (Gr), Brot-koch
*brouillés* (Fr), brou-il-lés
*broulaï* (Cb), brou-laï
*brown* (US), brown
*brown, to* (US), brown, to
*brown betty* (US), brown bet-ty
*brownie* (US), brown-ie
*brown rice* (US), brown rice
*brown sauce* (US), brown sauce
*brown stock* (US), brown stock
*brown sugar* (US), brown sug-ar
*brugnon* (Fr), bru-gnon
*Brühe* (Gr), Brühe
*bruin bonen* (Du), bruin bo-nen
*bruin brood* (Du), bruin brood

*brûlé* (Fr), brû-lé
*brûlot* (Fr), brû-lot
*bruna bönor* (Sw), bru-na bö-nor
*brunch* (US), brunch
*brunede kartofler* (Da), brune-de kar-to-fler
*brunekager* (Da), brune-ka-ger
*brun fisksuppe* (Nw), brun fisk-sup-pe
*brunkaalssupe* (Da), brun-kaal-sup-pe
*Brunnenkresse* (Gr), Brun-nen-kres-se
*brunoise* (Fr), bru-noise
*Brunswick stew* (US), Bruns-wick stew
*brus* (Nw), brus
*bruschetto* (It), bru-schet-to
*brush, to* (US), brush, to
*Brussels lof* (Du), Brus-sels lof
*Brussels sprouts* (US), Brus-sels sprouts
*Bruststück* (Gr), Brust-stück
*brut* (Fr), brut
*bruxelloise, à la* (Fr), bru-xell-oise, à la
*brylépudding* (Sw), bry-lé-pud-ding
*bryndza* (Po), bryn-dza
*brynt potatis* (Sw), brynt po-ta-tis
*brynza* (Cz), bryn-za
*brysselkål* (Sw), brys-sel-kål
*brzlík* (Cz), brz-lík
*brzoskwinia* (Po), br-zosk-wi-nia

*buah anggur* (In), bu-ah ang-gur
*buah ara* (In), bu-ah a-ra
*buah ceri* (In), bu-ah ce-ri
*buah zaitun* (In), bu-ah zai-tun
*bubble and squeak* (GB), bub-ble and squeak
*bubbly jock* (Sc), bub-bly jock
*bubrezi* (SC), bu-bre-zi
*bucatini* (It), bu-ca-ti-ni
*buccellato* (It), buc-cel-la-to
*buccin* (Fr), buc-cin
*bûche* (Fr), bûche
*bûche de Noël* (Fr), bûche de No-ël
*buchta* (Cz), buch-ta
*Buchweizen* (Gr), Buch-wei-zen
*Bückling* (Gr), Bück-ling
*buckwheat* (US), buck-wheat
*buckwheat flour* (US), buck-wheat flour
*budding* (Da), bud-ding
*budin* (Sp), bu-din
*bù-ding* (Ch), bù-ding
*budino* (It), bu-di-no
*budino di pasta* (It), bu-di-no di pa-sta
*budino di ricotta* (It), bu-di-no di ri-cot-ta
*budō* (Jp), bu-do
*budyń* (Po), bu-dyń
*bue* (It), bue
*buey* (Sp), buey
*Buffalo chicken wings* (US), Buf-fa-lo chick-en wings
*buffet* (Fr), buf-fet
*buggyantott tojás* (Hu), bug-gyan-tott to-jás

*buisson* (Fr), buis-son
*bulgar saláta* (Hu), bul-gar sa-lá-ta
*bulgur* (Ar), bul-gur
*buljong* (Sw), bul-jong
*bully beef* (GB), bul-ly beef
*buluchki* (Rs), bu-luch-ki
*bul'yon* (Rs), bul'yon
*bun* (GB), bun
*bun* (US), bun
*buncis* (In), bun-cis
*Bundnerfleisch* (Gr-Swiss), Bund-ner-fleisch
*buntil* (In), bun-til
*buñuelo* (Sp), bu-ñu-e-lo
*buñuelos de bacalla* (Sp), bu-ñu-e-los de ba-ca-lla
*buraki* (Po), bu-ra-ki
*buras* (In), bu-ras
*burbot* (GB), bur-bot
*burdock* (US), bur-dock
*burekakia* (Gk), bu-re-ka-kia
*burghul* (Ar), bur-ghul
*burgonya* (Hu), bur-gon-ya
*burgonyagombóc* (Hu), bur-go-nya-gom-bóc
*burgonyaleves* (Hu), bur-go-nya-le-ves
*burgoo* (US), bur-goo
*Burgos* (Sp), Bur-gos
*Burgundy* (Fr), Bur-gun-dy
*burnet* (US), burn-et
*burnt cream* (GB), burnt cream
*burritos* (Sp), bur-ri-tos
*burro* (It), bur-ro
*burro di acciuga* (It), bur-ro di ac-ciu-ga
*burro fuso* (It), bur-ro fu-so

*burro maggiordomo* (It), bur-ro mag-gior-do-mo
*burské ořísky* (Cz), bur-ské o-říš-ky
*buřtíky* (Cz), buř-tí-ky
*burtuaan* (Ar), bur-tu-aan
*busecca* (It), bu-sec-ca
*buta* (Jp), bu-ta
*butaniku* (Jp), bu-ta-ni-ku
*buta no kakuni* (Jp), bu-ta no ka-ku-ni
*butifarra* (Sp), bu-ti-far-ra
*butifarrón sabroso* (Cb), bu-ti-far-rón sa-bro-so
*Butt* (Gr), Butt
*butta* (Ar), but-ta
*butter* (US), but-ter
*Butter* (Gr), But-ter
*butter bean* (US), but-ter bean
*buttercream* (US), but-ter-cream
*buttercup squash* (US), but-ter-cup squash
*butterfat* (US), but-ter-fat
*butterfish* (US), but-ter-fish
*butterfly cut* (US), but-ter-fly cut
*Butterkäse* (Gr), But-ter-kä-se
*Buttermilchquark* (Gr), But-ter-milch-quark
*buttermilk* (US), but-ter-milk
*butternut* (US), but-ter-nut
*butternut squash* (US), but-ter-nut squash
*butterscotch* (US), but-ter-scotch

***Butterteig*** (Gr), But-ter-teig
***button mushroom*** (US), but-ton mush-room
***butyric acid*** (US), bu-ty-ric ac-id
***buz*** (Tr), buz

***buzhenina*** (Rs), bu-zhe-ni-na
***bwó káu-bing*** (Ch), bwó káu-bing
***bwō-lwó*** (Ch), bwō-lwó
***bwō-tsài*** (Ch), bwō-tsài
***bygg*** (Nw), bygg

# C

***caballa*** (Sp), ca-ba-lla
***cabbage*** (US), cab-bage
***cabbage palm*** (US), cab-bage palm
***cabe*** (In), ca-be
***Cabécou*** (Fr), Ca-bé-cou
***cabellos de ángel*** (Sp), ca-be-llos de án-gel
***cabe rawit*** (In), ca-be ra-wit
***cabillaud*** (Fr), ca-bi-llaud
***cabinet pudding*** (GB), cab-i-net pud-ding
***Caboc*** (Sc), Cab-oc
***cabra*** (Sp, Pg), ca-bra
***Cabrales*** (Sp), Ca-bra-les
***cabrito*** (Sp, Pg), ca-bri-to
***cabrito al horno*** (Sp), ca-bri-to al hor-no
***caça*** (Pg), ca-ça
***cacao*** (Du, Fr, It, Sp), ca-ca-o
***cacao*** (US), ca-ca-o
***cacau*** (Pg), ca-ca-u
***cacciagione*** (It), cac-cia-gio-ne

***cacciatora, alla*** (It), cac-cia-to-ra, al-la
***cacciucco*** (It), cac-ciuc-co
***cacerola*** (Sp), ca-ce-ro-la
***cacerola de pollo y elote*** (Mx), ca-ce-ro-la de po-llo y e-lo-te
***cachaca*** (Pg), ca-cha-ca
***cachorro*** (Pg), ca-cho-rro
***cachumbar*** (Ia), ca-chum-bar
***cacik*** (Tr), ca-cik
***cacio*** (It), ca-cio
***Caciocavallo*** (It), Ca-cio-ca-val-lo
***cacio grattato*** (It), ca-cio grat-ta-to
***cactus pear*** (US), cac-tus pear
***Caerphilly*** (GB), Caer-phil-ly
***Caesar salad*** (US), Cae-sar sal-ad
***Caesar's mushroom*** (US), Cae-sar's mush-room
***café*** (Fr, Sp, Pg), ca-fé
***café au lait*** (Fr), ca-fé au lait

*café brûlot* (Fr), ca-fé brû-lot

*café com crème e açúcar* (Pg), ca-fé com crème e a-çú-car

*café com leite* (Pg), ca-fé com le-i-te

*café complet* (Fr), ca-fé com-plet

*café con leche* (Sp), ca-fé con le-che

*café cortado* (Sp), ca-fé cor-ta-do

*café corto* (Sp), ca-fé cor-to

*cafedaki* (Gk), ca-fe-da-ki

*café express* (Fr), ca-fé ex-press

*café filtre* (Fr), ca-fé fil-tre

*café gelado* (Pg), ca-fé ge-la-do

*café glacé* (Fr), ca-fé gla-cé

*café noir* (Fr), ca-fé noir

*cafézinho* (Pg), ca-fé-zi-nho

*caffè cappuccino* (It), caf-fè cap-pu-ccino

*caffè espresso* (It), caf-fè es-pres-so

*caffeine* (US), caf-feine

*caffè latte* (It), caf-fè lat-te

*caffè nero* (It), caf-fè ne-ro

*cahn tom chua cai* (Vt), cahn tom chua cai

*caille* (Fr), caille

*caillebotte* (Fr), caille-botte

*caille-lait* (Fr), caille-lait

*cailles en sarcophage* (Fr), cailles en sar-co-phage

*čaj* (Cz, SC), ča-j

*cajeta* (Sp), ca-je-ta

*čaj s ledem* (Cz), ča-j s le-dem

*cajú* (Pg), ca-jú

*Cajun* (US), Ca-jun

*cake* (US), cake

*cake* (Du), cake

*calabash* (US), cal-a-bash

*calabaza* (Sp), ca-la-ba-za

*calamares en su tinta* (Sp), ca-la-ma-res en su tin-ta

*calamaretti* (It), cal-a-mar-et-ti

*calamari* (It), cal-a-mar-i

*calamari affogati* (It), cal-a-mar-i af-fo-ga-ti

*calamari imbottita* (It), cal-a-mar-i im-bot-ti-ta

*calamondin* (US), cal-a-mon-din

*calcium* (US), cal-ci-um

*calda de papa* (Sp), cal-da de pa-pa

*caldeirada* (Pg), cal-de-i-ra-da

*caldo* (It), cal-do

*caldo* (Pg, Sp), cal-do

*caldo de patas* (Sp), cal-do de pa-tas

*caldo de pescado* (Sp), cal-do de pes-ca-do

*caldo de pimentón* (Sp), cal-do de pi-men-tón

*caldo gallego* (It), cal-do gall-e-go

*caldo gallina* (It), cal-do gall-i-na

*caldo verde* (Pg), cal-do ver-de

*caléndula* (Sp), ca-lén-du-la

*calico scallops* (US), cal-i-co scal-lops

*cálido* (Sp), cá-li-do

*çali fasulye* (Tr), ça-li fa-sul-ye

*California green chili* (US), Cal-i-for-ni-a green chil-i

*Calimyrna fig* (US), Cal-i-myr-na fig

*callaloo* (Cb), cal-la-loo

*callaloo soup* (Cb), cal-la-loo soup

*callos* (Pg, Sp), ca-llos

*callos a la madrileña* (Sp), ca-llos a la ma-dri-le-ña

*callos de porco com amêijoas* (Pg), ca-llos de por-co com a-mê-i-jo-as

*calmar* (Fr), cal-mar

*calorie* (US), cal-o-rie

*calorie-free* (US), cal-o-rie-free

*calvados* (Fr), cal-va-dos

*calzone* (It), cal-zone

*camarão* (Pg), ca-ma-rã-o

*camarones* (Sp), ca-ma-ro-nes

*Camembert* (Fr), Cam-em-bert

*camote* (Sp), ca-mo-te

*campagnola, alla* (It), cam-pa-gno-la, al-la

*Canadian bacon* (Ca), Can-a-di-an ba-con

*canalons* (Sp), ca-na-lons

*canapé* (Fr), can-a-pé

*canard* (Fr), ca-nard

*canard à l'orange* (Fr), ca-nard à l'or-ange

*canard aux cerises* (Fr), ca-nard aux ce-ri-ses

*canard rôti* (Fr), ca-nard rô-ti

*canard sauvage* (Fr), ca-nard sau-vage

*candát* (Cz), can-dát

*candito* (It), can-di-to

*candy* (US), can-dy

*candy, to* (US), can-dy, to

*canela* (Sp), ca-ne-la

*cane syrup* (US), cane syr-up

*caneton* (Fr), can-e-ton

*cangrejo* (Sp), can-gre-jo

*cangrejo de rio* (Sp), can-gre-jo de ri-o

*canja* (Pg), can-ja

*canneberge* (Fr), can-ne-berge

*cannella* (It), can-nel-la

*cannelle* (Fr), can-nelle

*cannellini* (It), can-nel-li-ni

*cannelloni* (It), can-nel-lo-ni

*canning* (US), can-ning

*cannoli* (It), can-no-li

*canola oil* (US), ca-no-la oil

*Cantal* (Fr), Can-tal

*cantaloupe* (US), can-ta-loupe

*cantarello* (It), can-ta-rel-lo

*canterellen* (Du), can-te-rel-len

*cap cay* (In), cap cay

*capellini* (It), ca-pel-li-ni

*capercailzie* (Sc), cap-er-cail-zie

*capers* (US), ca-pers

*capirotada* (Mx), ca-pi-ro-ta-da

*capitone* (It), ca-pi-to-ne

*capocollo* (It), ca-po-col-lo

*capon* (US), ca-pon

*caponatina* (It), ca-po-na-ti-na

*cappelletti* (It), cap-pel-let-ti

*cappelli d'angelo* (It), cap-pel-li d'an-ge-lo

*capperi* (It), cap-per-i

*cappesante* (It), cap-pe-san-te

*cappone* (It), cap-po-ne

*cappuccino* (It), cap-pu-cci-no

*câpres, sauce aux* (Fr), câ-pres, sauce aux

*capretto* (It), ca-pret-to

*capsicum* (US), cap-si-cum

*capuchino* (Sp), ca-pu-chi-no

*caracóis* (Pg), ca-ra-cóis

*caracoles en salsa* (Sp), ca-ra-co-les en sal-sa

*carambola* (US), car-am-bo-la

*caramel* (US), car-a-mel

*caramelize, to* (US), car-a-mel-ize, to

*caramelle* (It), ca-ra-melle

*caranguejo* (Pg), car-an-gue-jo

*caraway seeds* (US), car-a-way seeds

*carbohydrate* (US), car-bo-hy-drate

*carbonara, alla* (It), car-bo-na-ra, al-la

*carbonated water* (US), car-bon-at-ed wa-ter

*carbonnade à la flamande* (Fr), car-bon-nade à la flam-ande

*carbonnades de boeuf* (Fr), car-bon-nades de boeuf

*carcakes* (Sc), car-cakes

*carciofi* (It), car-ci-o-fi

*carciofini all'olio* (It), car-ci-o-fi-ni al-l'o-lio

*cardamom* (US), car-da-mom

*carde* (Fr), carde

*cardon* (Fr), car-don

*cardoon* (US), car-doon

*cari* (Fr), ca-ri

*carne* (It, Pg, Sp), car-ne

*carne al sange* (It), car-ne al sange

*carne asada* (Sp), car-ne a-sa-da

*carne ben cotta* (It), car-ne ben cot-ta

*carne de membrillo* (Sp), car-ne de mem-bri-llo

*carne de res* (Mx), car-ne de res

*carne de venado* (Sp), car-ne de ve-na-do

*carne de vinho e albos* (Pg), car-ne de vi-nho e a-lhos

*carneiro* (Pg), car-ne-i-ro

*carne non troppo cotto* (It), car-ne non trop-po cot-to

*carnitas* (Mx), car-ni-tas

*carob* (US), car-ob

*carote* (It), ca-ro-te

*carotene* (US), car-o-tene

*carottes* (Fr), ca-rottes

*carp* (US), carp

*carpe* (Fr), carpe

*carpio* (It), car-pio

*carrageenin* (US), car-ra-gee-nin

*carré d'agneau aux herbes* (Fr), car-ré d'agn-eau aux herbes

*carré de porc provençal*

(Fr), car-ré de porc pro-ven-çal

**carrelet** (Fr), car-re-let

**carrot** (US), car-rot

**carrot cake** (US), car-rot cake

**casaba** (US), ca-sa-ba

**casalinga** (It), ca-sa-lin-ga

**cascabel** (US), cas-ca-bel

**casein** (US), ca-sein

**cashew** (US), cash-ew

**cassareep** (Cb), cas-sa-reep

**cassata** (It), cas-sa-ta

**cassava** (US), cas-sa-va

**casserole** (US), cas-se-role

**cassia** (US), cas-sia

**cassis** (Fr), cas-sis

**cassoulet** (Fr), cas-sou-let

**cassoulet toulousain** (Fr), cas-sou-let tou-lou-sain

**castagna** (It), cas-ta-gna

**castagnaccio** (It), cas-ta-gna-ccio

**castañas con jaraba** (Sp), cas-ta-ñas con ja-ra-ba

**castanha** (Pg), cas-ta-nha

**castanha-do-pará** (Pg), cas-ta-nha-do-pa-rá

**caster sugar** (GB), cas-ter sug-ar

**catfish** (US), cat-fish

**cat's tongue** (US), cat's tongue

**catsup** (US), cat-sup

**caul** (US), caul

**cauliflower** (US), cau-li-flow-er

**cavatelli** (It), ca-va-tel-li

**caviale** (It), ca-vi-a-le

**caviar** (US), cav-i-ar

**cavolfiore** (It), ca-vol-fio-re

**cavoli** (It), ca-vo-li

**cavoli imbottiti** (It), ca-vo-li im-bot-ti-ti

**cavolino di Brusselle** (It), ca-vo-li-no di Brus-selle

**cavolo riccio** (It), ca-vo-lo ri-ccio

**cavolo rosso** (It), ca-vo-lo ros-so

**cavolrapa** (It), ca-vol-ra-pa

**çay** (Tr), çay

**cayenne pepper** (US), cay-enne pep-per

**cazabe** (Sp), ca-za-be

**cebolas** (Pg), ce-bo-las

**cebollas** (Sp), ce-bo-llas

**cebula** (Po), ce-bu-la

**ceci** (It), ce-ci

**cecina** (Sp), ce-ci-na

**cefalo** (It), ce-fa-lo

**céklasaláta** (Hu), cék-la-sa-lá-ta

**celer** (SC, Cz), ce-ler

**céleri** (Fr), cé-le-ri

**celeriac** (US), ce-ler-i-ac

**céleri-rave rémoulade** (Fr), cé-le-ri-rave ré-mou-lade

**celery** (US), cel-er-y

**celery salt** (US), cel-er-y salt

**celery seeds** (US), cel-er-y seeds

**cellophane noodles** (US), cel-lo-phane noo-dles

**cendawan** (In), cen-da-wan

**cenouras** (Pg), ce-nou-ras

**centeno** (Sp), cen-te-no

**cèpe** (Fr), cèpe

**cèpes farcis** (Fr), cèpes far-cis

**ceppatella** (It), cep-pa-tel-la

*cereale cotto* (It), ce-re-a-le
cot-to
*céréales* (Fr), cé-ré-ales
*cerejas* (Pg), ce-re-jas
*cereza* (Sp), ce-re-za
*çerezler* (Tr), çe-rez-ler
*cerfeuil* (Fr), cer-feuil
*cerfoglio* (It), cer-fo-gli-o
*cerise* (Fr), ce-rise
*çerkez tavuğu* (Tr), çer-kez
ta-vuğu
*čerstvé* (Cz), čerst-vé
*cerveja* (Pg), cer-ve-ja
*cervelles* (Fr), cer-velles
*cervelles au beurre noir*
(Fr), cer-velles au beurre
noir
*cervelli fritti* (It), cer-vel-li
frit-ti
*červená řepa* (Cz), čer-ve-ná
ře-pa
*červené vino* (Cz), čer-ve-né
vi-no
*cerveza* (Sp), cer-ve-za
*česnek* (Cz), čes-nek
*cetriolo* (It), ce-tri-o-lo
*ćevapčići* (SC), će-vap-či-ći
*ceviche* (Sp), ce-vi-che
*chá* (Ch), chá
*chá* (Pg), chá
*Chablis* (Fr), Cha-blis
*chá com limão* (Pg), chá com
li-mã-o
*chá gelado* (Pg), chá ge-la-do
*cha gio* (Vt), cha gio
*chai* (Af-Swahili), cha-i
*chai* (Rs), chai
*chakchouka* (Ar), chak-chou-
ka

*chakleti* (Af-Swahili), cha-
kle-ti
*chakula* (Af-Swahili), cha-ku-
la
*chalky* (US), chalk-y
*challah* (Jw), chal-lah
*chamomile tea* (US), cham-
o-mile tea
*champ* (Ir), champ
*champagne* (Fr), cham-pa-
gne
*champagne, au* (Fr), cham-
pa-gne, au
*champanhe* (Pg), cham-pa-
nhe
*champignons* (Fr), cham-pi-
gnons
*champignons* (Du), cham-
pign-ons
*champignons à blanc* (Fr),
cham-pi-gnons à blanc
*champignons à la grecque*
(Fr), cham-pi-gnons à la
grecque
*champignons, aux* (Fr),
cham-pi-gnons, aux
*champignons farcis* (Fr),
cham-pi-gnons far-cis
*champiñones* (Sp), cham-pi-
ño-nes
*champurrado* (Sp), cham-
pu-rra-do
*channa* (Ia), chan-na
*chanoki* (Rs), cha-no-ki
*chanterelle* (Fr), chan-te-
relle
*Chantilly* (Fr), Chan-til-ly
*chapati* (Ia), cha-pa-ti
*chapon* (Fr), cha-pon

**chapon de Gascogne** (Fr), cha-pon de Gas-cogne

**char** (US), char

**charcuterie** (Fr), char-cu-te-rie

**chard** (US), chard

**charlotte** (Fr), char-lotte

**charlotte russe** (Fr), char-lotte russe

**chartreuse** (Fr), char-treuse

**chă shău** (Ch), chă shău

**chashmé ka paani** (Ia), chash-mé ka paan-i

**cha soba** (Jp), cha so-ba

**châtaigne** (Fr), châ-tai-gne

**chateaubriand** (Fr, Nw), cha-teau-bri-and

**chatini** (Af-Swahili), cha-ti-ni

**chău bái-tsài** (Ch), chău bái-tsài

**chaud-froid** (Fr), chaud-froid

**chău dòu-fú** (Ch), chău dòu-fú

**chău-fán** (Ch), chău-fán

**chău ji-dàn** (Ch), chău ji-dàn

**chău-myàn** (Ch), chău-myàn

**chausson** (Fr), chaus-son

**chău syā-rén** (Ch), chău syā-rén

**chaval** (Ia), cha-val

**chawan mushi** (Jp), cha-wan mu-shi

**chayote** (US), cha-yo-te

**chaza** (Af-Swahili), cha-za

**chebureki** (Rs), che-bu-re-ki

**Cheddar** (US), Ched-dar

**cheese** (US), cheese

**cheesecake** (US), cheese-cake

**chelo** (Ar), che-lo

**chenza** (Af-Swahili), che-nza

**cherimoya** (Sp), che-ri-moy-a

**cherry** (US), cher-ry

**cherrystone clam** (US), cher-ry-stone clam

**chervil** (US), cher-vil

**chess pie** (US), chess pie

**chestnut** (US), chest-nut

**cheval, à** (Fr), che-val, à

**chèvre** (Fr), chèvre

**Chèvre** (Fr), Chèvre

**chewa** (Af-Swahili), che-wa

**chianti** (It), chi-an-ti

**Chiboust** (Fr), Chi-boust

**chichinda** (Ia), chi-chin-da

**chicken** (US), chick-en

**chicken à la King** (US), chick-en à la King

**chicken-fried steak** (US), chick-en-fried steak

**chicken Kiev** (US), chick-en Ki-ev

**chicken Tetrazzini** (US), chick-en Te-traz-zi-ni

**chick-pea** (US), chick-pea

**chicorée** (Fr), chi-co-rée

**chicory** (US), chic-o-ry

**chikuwa** (Jp), chi-ku-wa

**chile** (Mx, Sp), chi-le

**chiles en nogada** (Sp), chi-les en no-ga-da

**chiles rellenos** (Sp), chi-les rel-le-nos

**chili con carne** (US), chil-i con car-ne

**chili con queso** (US), chil-i con que-so

**chili paste** (US), chil-i paste

*chili pepper* (US), chil-i pep-per

*chili powder* (US), chil-i pow-der

*chimichangas* (Mx), chi-mi-chan-gas

*Chinese anise* (US), Chi-nese an-ise

*Chinese artichoke* (US), Chi-nese ar-ti-choke

*Chinese black vinegar* (US), Chi-nese black vin-e-gar

*Chinese cabbage* (US), Chi-nese cab-bage

*Chinese gooseberry* (US), Chi-nese goose-ber-ry

*Chinese parsley* (US), Chi-nese pars-ley

*Chinese sausage* (US), Chi-nese sau-sage

*chīng-jyāu* (Ch), chīng-jyāu

*chīng tāng* (Ch), chīng tāng

*chīng-yú* (Ch), chīng-yú

*chín-jyāu* (Ch), chín-jyāu

*chinook salmon* (US), chi-nook sal-mon

*chín-tsài* (Ch), chín-tsài

*chiòdo di garofano* (It), chi-ò-do di ga-ro-fa-no

*chipolata* (Fr), chi-po-la-ta

*chipotle chile* (Mx), chi-po-tle chi-le

*chipped beef* (US), chip-ped beef

*chirashizushi* (Jp), chi-ra-shi-zu-shi

*chiri mushi* (Jp), chi-ri mu-shi

*chirinabe* (Jp), chi-ri-na-be

*chispe* (Pg), chis-pe

*chitterlings* (US), chit-ter-lings

*chitlins* (US), chit-lins

*chive* (US), chive

*Chivry* (Fr), Chi-vry

*chí-yú* (Ch), chí-yú

*chléb* (Cz), chléb

*chlodnik z ryby* (Po), chlod-nik z ry-by

*chocola* (Du), cho-co-la

*chocolate* (US), choc-o-late

*chocolate quente* (Pg), cho-co-la-te quen-te

*choix* (Fr), choix

*chokladglass* (Sw), chok-lad-glass

*cholent* (Jw), cho-lent

*cholesterol* (US), cho-les-ter-ol

*cholesterol-free* (US), cho-les-ter-ol-free

*chop, to* (US), chop, to

*chop suey* (US), chop su-ey

*chorizo* (Mx), chor-i-zo

*chorogi* (Jp), cho-ro-gi

*chou* (Fr), chou

*chou-broccoli* (Fr), chou-broc-co-li

*choucroute* (Fr), chou-croute

*choucroute garnie* (Fr), chou-croute gar-nie

*chou de Milan* (Fr), chou de Mi-lan

*chou-fleur* (Fr), chou-fleur

*chou frisé* (Fr), chou fri-sé

*chou-rave* (Fr), chou-rave

*chouriço* (Pg), chou-ri-ço

*chou rouge* (Fr), chou rouge

*chou vert* (Fr), chou vert

*choux* (Fr), choux

**choux à la crème** (Fr), choux à la crème

**choux de Bruxelles** (Fr), choux de Bru-xelles

**choux de Bruxelles à la grandmère** (Fr), choux de Bru-xelles à la grand-mère

**choux pastry** (Fr), choux past-ry

**chow chow** (US), chow chow

**chowder** (US), chow-der

**chowder clam** (US), chow-der clam

**chow mein** (US), chow mein

**chřest** (Cz), chřest

**chromium** (US), chro-mi-um

**chrysanthemum** (US), chry-san-the-mum

**chrzan** (Po), chr-zan

**chub** (US), chub

**chuletas de cordero** (Sp), chu-le-tas de cor-de-ro

**chuletas de res** (Sp), chu-le-tas de res

**chuletas de ternera** (Sp), chu-le-tas de ter-ne-ra

**chumvi** (Af-Swahili), chu-mvi

**chungurro** (Sp), chun-gu-rro

**chungwa** (Af-Swahili), chu-ngwa

**churrasco** (Pg, Sp), chu-rras-co

**churros** (Sp), chu-rros

**chutney** (Ia), chut-ney

**chwūn-jywǎn** (Ch), chwūn-jywǎn

**chyáu-mài** (Ch), chyáu-mài

**chyé-dz** (Ch), chyé-dz

**chyōu-kwèi** (Ch), chyōu-kwèi

**chywán-shú de jī-dàn** (Ch), chywán-shú de jī-dàn

**chywè-mài** (Ch), chywè-mài

**cialda** (It), ci-al-da

**ciambelle** (It), ci-am-bel-le

**ciasteczka** (Po), cias-tecz-ka

**ciastka drozdzowe** (Po), cia-st-ka droz-dzo-we

**ciastka miodowe** (Po), ciast-ka mio-do-we

**ciasto** (Po), cias-to

**ciboulette** (Fr), ci-bou-lette

**cibrèo** (It), ci-brè-o

**cibule** (Cz), ci-bu-le

**cicoria** (It), ci-co-ri-a

**cider** (US), ci-der

**cielecina** (Po), cie-le-ci-na

**cigala** (Sp), ci-ga-la

**cigány gulyás** (Hu), ci-gány gu-lyás

**ciğer** (Tr), ciğ-er

**çikolata** (Tr), çi-ko-la-ta

**cilantrillo** (Sp-Puerto Rico), ci-lan-tri-llo

**cilantro** (Sp), ci-lan-tro

**çilek** (Tr), çi-lek

**çilek-kaymali** (Tr), çi-lek-kay-ma-li

**ciliege** (It), ci-lie-ge

**cima di vitello** (It), ci-ma di vi-tel-lo

**cimino** (It), ci-mi-no

**cinnamon** (US), cin-na-mon

**cioccolata calda** (It), cioc-co-la-ta cal-da

**cioppino** (US), ciop-pi-no

**ciorba** (Ru), cior-ba

**cipolle** (It), ci-pol-le

**cipolline** (It), cip-ol-li-ne

**cipolline in agrodolce** (It), cip-ol-li-ne in ag-ro-dol-ce
**ciruela** (Sp), ci-ru-e-la
**citric acid** (US), cit-ric ac-id
**citroen** (Du), ci-tro-en
**citron** (Fr), ci-tron
**citron** (US), cit-ron
**citrouille** (Fr), ci-trou-ille
**citrus** (US), cit-rus
**civet** (Fr), civ-et
**civet de lièvre** (Fr), ci-vet de lièvre
**clabber** (US), clab-ber
**clafouti** (Fr), cla-fou-ti
**clam** (US), clam
**Clamart** (Fr), Cla-mart
**clambake** (US), clam-bake
**clam chowder** (US), clam chow-der
**clarified butter** (US), clar-i-fied but-ter
**clarify** (US), clar-i-fy
**clavo** (Sp), cla-vo
**clotted cream** (GB), clot-ted cream
**cloudberry** (Ca), cloud-ber-ry
**cloud ear** (US), cloud ear
**clou de girofle** (Fr), clou de gi-ro-fle
**cloves** (US), cloves
**coat, to** (US), coat, to
**cobbler** (US), cob-bler
**cochifrito** (Sp), co-chi-fri-to
**cochinillo asado** (Sp), co-chi-ni-llo a-sa-do
**cochon** (Fr), co-chon
**cochon au lait** (Fr), co-chon au lait
**cocido** (Sp), co-ci-do

**cocido al vapor** (Sp), co-ci-do al va-por
**cocido de riñones** (Sp), co-ci-do de ri-ño-nes
**čočka** (Cz), čoč-ka
**cock-a-leekie** (Sc), cock-a-leek-ie
**čočková polévka** (Cz), čoč-ko-vá po-lév-ka
**coco** (Sp), co-co
**cocoa** (US), co-coa
**cocoa bean** (US), co-coa bean
**cocomero** (It), co-co-me-ro
**coconut** (US), co-co-nut
**coconut milk** (US), co-co-nut milk
**coco quemado** (Sp), co-co que-ma-do
**cocotte, en** (Fr), co-cotte, en
**coctel de mariscos** (Sp), coc-tel de ma-ris-cos
**cod** (US), cod
**coddle, to** (US), cod-dle, to
**codorniz** (Sp), co-dor-niz
**coêlho** (Pg), co-ê-lho
**coeur à la crème** (Fr), coeur à la crème
**coeur d'artichauts** (Fr), coeur d'ar-ti-chauts
**coeur de laitue** (Fr), coeur de lai-tue
**coeur de palmier** (Fr), coeur de palm-i-er
**coffee** (US), cof-fee
**cognac** (Fr), co-gnac
**cogumelos** (Pg), co-gu-me-los
**coing** (Fr), coing
**çok** (Tr), çok

*col* (Sp), col
*Colby* (US), Col-by
*colcannon* (Ir), col-can-non
*cold cuts* (US), cold cuts
*col de Bruselas* (Sp), col de Bru-se-las
*cold pack* (US), cold pack
*coleslaw* (US), cole-slaw
*coliflor* (Sp), co-li-flor
*coliflor al ajo arriero* (Sp), co-li-flor al a-jo ar-ri-e-ro
*colin* (Fr), co-lin
*collards* (US), col-lards
*col lombada* (Sp), col lom-ba-da
*collop* (GB), col-lop
*col rizada* (Sp), col ri-za-da
*combine* (US), com-bine
*comida* (Pg), co-mi-da
*comino* (Mx), co-mi-no
*common edible crab* (GB), com-mon ed-i-ble crab
*complete protein* (US), com-plete pro-tein
*compota de manzana* (Sp), com-po-ta de man-za-na
*compote de fruits* (Fr), com-pote de fruits
*compound butter* (US), com-pound but-ter
*compressed yeast* (US), com-pressed yeast
*con* (It), con
*concassé* (Fr), con-cas-sé
*conch* (US), conch
*conchiglie* (It), con-chi-glie
*concombres* (Fr), con-com-bres
*condensed milk* (US), con-densed milk

*condimentos* (Pg), con-di-men-tos
*conejo* (Sp), co-ne-jo
*confectioners' custard* (US), con-fec-tion-ers' cus-tard
*confectioners' sugar* (US), con-fec-tion-ers' sug-ar
*confit* (Fr), con-fit
*confit d'oie* (Fr), con-fit d'oie
*confiture* (Fr), con-fi-ture
*congee* (Ch), con-gee
*congee* (Ia), con-gee
*congelato* (It), con-ge-la-to
*congelé* (Fr), con-ge-lé
*coniglio* (It), co-ni-glio
*conserva* (Sp), con-ser-va
*conserve* (US), con-serve
*conserve au vinaigre* (Fr), con-serve au vin-ai-gre
*consommé* (Fr), con-som-mé
*consommé à l'alsacienne* (Fr), con-som-mé à l'al-sa-cienne
*consommé à la bourgeoise* (Fr), con-som-mé à la bourge-oise
*consommé à la madrilène* (Fr), con-som-mé à la ma-dri-lène
*consommé à la reine* (Fr), con-som-mé à la reine
*consommé aux perles* (Fr), con-som-mé aux perles
*consommé doré* (Fr-Creole), con-som-mé do-ré
*consommé froid* (Fr), con-som-mé froid
*consommé julienne* (Fr), con-som-mé ju-lienne

*consommé printanier* (Fr), con-som-mé prin-tan-ier

*consommé queue de boeuf* (Fr), con-som-mé queue de boeuf

*Conti* (Fr), Con-ti

*contorno* (It), con-tor-no

*contre-filet* (Fr), con-tre-fi-let

*cookie* (US), cook-ie

*cookie* (Sc), cook-ie

*Coon Cheddar* (US), Coon Ched-dar

*coppa* (It), cop-pa

*copper* (US), cop-per

*coq au vin* (Fr), coq au vin

*coque, à la* (Fr), coque, à la

*coquillages* (Fr), co-quil-lages

*coquilles St. Jacques* (Fr), co-quilles St. Jacques

*coquilles St. Jacques à la parisienne* (Fr), co-quilles St. Jacques à la par-is-ienne

*çorba* (Tr), çor-ba

*čorba od povrća* (SC), čor-ba od po-vr-ća

*corbina y mariscos al vapor* (Sp), cor-bi-na y ma-ris-cos al va-por

*cordeiro* (Pg), cor-de-i-ro

*cordon bleu* (Fr), cor-don bleu

*core, to* (US), core, to

*coriander* (US), co-ri-an-der

*corn* (US), corn

*corn bread* (US), corn bread

*corn dog* (US), corn dog

*corned beef* (US), corned beef

*cornet* (Fr), cor-net

*cornflour* (GB), corn-flour

*cornichon* (Fr), cor-ni-chon

*Cornish pasty* (GB), Corn-ish pas-ty

*corn meal* (US), corn meal

*corn oil* (US), corn oil

*corn salad* (US), corn sal-ad

*cornstarch* (US), corn-starch

*corn syrup* (US), corn syr-up

*còscia* (It), còs-cia

*còscia di agnello arrosto* (It), còs-cia di a-gnel-lo ar-ros-to

*cos lettuce* (GB), cos let-tuce

*costelêta* (Pg), cos-te-lê-ta

*costelêta de porco* (Pg), cos-te-lê-ta de por-co

*costillas de cerdo* (Sp), cos-ti-llas de cer-do

*costmary* (GB), cost-mary

*costole* (It), cos-to-le

*costolette* (It), cos-to-let-te

*costolette alla milanese* (It), cos-to-let-te al-la mi-la-ne-se

*costolette alla Modenese* (It), cos-to-let-te al-la Mo-de-ne-se

*costolette di agnello piccante* (It), cos-to-let-te di a-gnel-lo pic-can-te

*costolette di maiale* (It), cos-to-let-te di ma-ia-le

*cotechino* (It), co-te-chi-no

*côte de boeuf grillé* (Fr), côte de boeuf gril-lé

*côtelette d'agneau* (Fr), côte-lette d'ag-neau

*côtelette de porc* (Fr), côte-lette de porc

*côtelette de veau* (Fr), côte-lette de veau
*cotignac* (Fr), co-ti-gnac
*cotognata* (It), co-to-gna-ta
*cotolette* (It), co-to-let-te
*cotolette alla milanese* (It), co-to-let-te al-la mi-la-ne-se
*cottage cheese* (US), cot-tage cheese
*cottage pudding* (US), cot-tage pud-ding
*cotto* (It), cot-to
*cotto a vapore* (It), cot-to a va-po-re
*cotton candy* (US), cot-ton can-dy
*cottonseed oil* (US), cot-ton-seed oil
*coulibiac de saumon de croûte* (Fr), cou-li-biac de sau-mon de croûte
*coulis* (Fr), cou-lis
*coupe aux marrons* (Fr), coupe aux mar-rons
*coupe de fruits frais* (Fr), coupe de fruits frais
*courgette* (Fr), cour-gette
*court bouillon* (Fr), court bou-illon
*court bouillon à la Creole* (Fr), court bou-illon à la Cre-ole
*couve* (Pg), cou-ve
*couveflor* (Pg), cou-ve-flor
*cozida à portuguesa* (Pg), co-zi-da à por-tu-gue-sa
*cozido* (Pg), co-zi-do
*cozido no forno* (Pg), co-zi-do no for-no

*cozze* (It), coz-ze
*crab* (US), crab
*crabe* (Fr), crabe
*crab Louis* (US), crab Lou-is
*cracker* (US), crack-er
*crackling bread* (US), crack-ling bread
*cracklings* (US), crack-lings
*cracknel* (GB), crack-nel
*cranberry* (US), cran-ber-ry
*cranberry sauce* (US), cran-ber-ry sauce
*crapaudine, à la* (Fr), cra-pau-di-ne, à la
*crappin* (Sc), crap-pin
*crappit heids* (Sc), crap-pit heids
*crawdad* (US), craw-dad
*crawfish* (US), craw-fish
*crayfish* (US), cray-fish
*cream* (US), cream
*cream, to* (US), cream, to
*cream cheese* (US), cream cheese
*cream horn* (US), cream horn
*cream of tartar* (US), cream of tar-tar
*cream puff* (US), cream puff
*cream puff pastry* (US), cream puff past-ry
*Crécy, à la* (Fr), Cré-cy, à la
*crema* (It), cre-ma
*crema catalana* (Sp), cre-ma ca-ta-la-na
*Crema Dania* (Da), Cre-ma Dan-i-a
*crema pastelera al ron* (Sp), cre-ma pas-te-le-ra al ron

*crema pasticcera* (It), cre-ma pas-tic-cer-a

*crême* (Pg), crême

*crême* (Fr), crême

*crème anglaise* (Fr), crème ang-laise

*crème brûlée* (Fr), crème brû-lée

*crème caramel* (Fr), crème ca-ra-mel

*crème Chantilly* (Fr), crème Chan-til-ly

*crème d'asperges* (Fr), crème d'as-per-ges

*crème de tomates* (Fr), crème de to-mates

*crème de volaille* (Fr), crème de vo-laille

*crème fouettée* (Fr), crème fou-et-tée

*crème fraîche* (Fr), crème fraîche

*crème glacée* (Fr), crème gla-cée

*Creole* (US), Cre-ole

*créole, à la* (Fr), cré-ole, à la

*crêpe* (Fr), crêpe

*crêpes alsaciennes* (Fr), crêpes al-sa-ciennes

*crêpes de homard* (Fr), crêpes de ho-mard

*crêpes Suzette* (Fr), crêpes Su-zette

*crêpinette* (Fr), crê-pin-ette

*Crescenza* (It), Cres-cen-za

*crescione* (It), cres-cio-ne

*crespelle alla fiorentina* (It), cres-pel-le al-la fio-ren-ti-na

*cress* (US), cress

*cresson de ruisseau* (Fr), cres-son de ru-is-seau

*crevette* (Fr), cre-vette

*crevette rose* (Fr), cre-vette rose

*crimp, to* (US), crimp, to

*crisp* (US), crisp

*crisp* (GB), crisp

*crisp, to* (US), crisp, to

*crni bleb* (SC), cr-ni hleb

*crni luk* (SC), cr-ni luk

*crno vino* (SC), cr-no vi-no

*croccanti* (It), croc-can-ti

*crocchetta* (It), croc-chet-ta

*croissant* (Fr), crois-sant

*croque madame* (Fr), croque ma-dame

*croquembouche* (Fr), cro-quem-bouche

*croque monsieur* (Fr), croque mon-sieur

*croquette* (Fr), cro-quette

*crosnes* (Fr), cros-nes

*crostacei* (It), cros-ta-cei

*crostata* (It), cros-ta-ta

*crostini* (It), cros-ti-ni

*crostini en brodo* (It), cros-ti-ni en bro-do

*croustade* (Fr), crou-stade

*croûte, en* (Fr), croûte, en

*croûtons* (Fr), croû-tons

*crowdie* (Sc), crowd-ie

*Crowdie* (Sc), Crowd-ie

*cru* (Fr), cru

*crudités* (Fr), cru-di-tés

*crudo* (It, Sp), cru-do

*cruller* (US), crul-ler

*crumb, to* (US), crumb, to

*crumble* (GB), crum-ble

*crumpet* (GB), crum-pet
*crust* (US), crust
*csalamádé* (Hu), csa-la-má-dé
*cseresznye* (Hu), cse-resz-nye
*csiga* (Hu), csi-ga
*csipetke* (Hu), csi-pet-ke
*csirke* (Hu), csir-ke
*csirkeleves* (Hu), csir-ke-le-ves
*csuka* (Hu), csu-ka
*cube, to* (US), cube, to
*cucumber* (US), cu-cum-ber
*ćufte* (SC), ćuf-te
*cuicere alla graticola* (It), cui-ce-re al-la gra-ti-co-la
*cuisine* (Fr), cui-sine
*cuisine minceur* (Fr), cui-sine min-ceur
*cuisses de grenouilles* (Fr), cuis-ses de gre-nou-illes
*cuissot de porc* (Fr), cuis-sot de porc
*cuissot de marcassin* (Fr), cuis-sot de mar-cas-sin
*cuit* (Fr), cuit
*cuit au four* (Fr), cuit au four
*cukier* (Po), cu-kier
*cukor* (Hu), cu-kor
*cukr* (Cz), cu-kr
*cukrovi* (Cz), cu-kro-vi
*ćulbastija* (SC), ćul-ba-sti-ja

*cullen skink* (Sc), cul-len skink
*cumin* (US), cum-in
*cuore* (It), cuo-re
*cupcake* (US), cup-cake
*curd* (US), curd
*cure, to* (US), cure, to
*currant* (US), cur-rant
*curry* (US), cur-ry
*curry leaves* (US), cur-ry leaves
*curry powder* (US), cur-ry pow-der
*curry sauce* (GB), cur-ry sauce
*cuscinetti di vitello* (It), cus-ci-net-ti di vi-tel-lo
*custard* (US), cus-tard
*custard apple* (US), cus-tard ap-ple
*cut in* (US), cut in
*cutlet* (US), cut-let
*cuttlefish* (US), cut-tle-fish
*cwikla* (Po), cwik-la
*cyanocobalamin* (US), cy-an-o-co-bal-a-min
*cymling* (US), cym-ling
*cytryna* (Po), cyt-ry-na
*czekolada* (Po), cze-ko-la-da
*czosnek* (Po), czos-nek

# *D*

*dab* (US), dab

*dà-bi-mù-yú* (Ch), dà-bi-mù-
yú

*dacca* (Ia), dac-ca

*dacquoise* (Fr), dac-quoise

*dadar* (In), da-dar

*dadar kepiting* (In), da-dar
ke-pi-ting

*dadlar* (Sw), dad-lar

*dadler* (Nw), dad-ler

*dagaa* (Af-Swahili), da-ga-a

*daging* (In), da-ging

*daging masak djahe* (In),
da-ging ma-sak dja-he

*daging redang* (In), da-ging
re-dang

*daging rusa* (In), da-ging rusa

*daging sapi* (In), da-ging sa-
pi

*daging sapi cincang* (In),
da-ging sa-pi cin-cang

*daging sapi gulung* (In), da-
ging sa-pi gu-lung

*daging sapi panggang* (In),
da-ging sa-pi pang-gang

*dag kavush* (Jw), dag ka-vush

*dagnje* (SC), dag-nje

*dahi* (Ia), da-hi

*dahi alu puri* (Ia), da-hi a-lu
pu-ri

*dahi bara* (Ia), da-hi ba-ra

*dahi bath* (Ia), da-hi bath

*dà-hwáng* (Ch), dà-hwáng

*daikon* (Jp) dai-kon

*daizu* (Jp), dai-zu

*dajaj mahshi* (Ar), da-jaj mah-
shi

*dal* (Ia), dal

*dalchini* (Ia), dal-chi-ni

*dalia* (Ru), dal-ia

*dal papri* (Ia), dal pa-pri

*dà-mài* (Ch), dà-mài

*damascos* (Pg), da-mas-cos

*Dampfnudeln* (Gr), Dampf-
nu-deln

*damper* (Aal, damp-er

*Damson* (US), Dam-son

*Damwildkeule* (Gr), Dam-
wild-keu-le

*dana* (Tr), da-na

*Danbo* (Da), Dan-bo

*dàn-bù-dīng* (Ch), dàn-bù-
dīng

*dandelion* (US), dan-de-li-on

*dandelion greens* (US), dan-
de-li-on greens

*dango* (Jp), dan-go

*dàn hwā tāng* (Ch), dàn hwā tāng

*Danish Blue* (Da), Dan-ish Blue

*Danish pastry* (US), Dan-ish past-ry

*Danska weinerbrød* (Da), Dan-ska wei-ner-brød

*Dansk leverpostej* (Da), Dansk le-ver-pos-tej

*daragaluska* (Hu), da-ra-ga-lus-ka

*darált marhahús* (Hu), da-rált mar-ha-hús

*dariole* (Fr), dar-i-ole

*Darjeeling* (Ia), Dar-jee-ling

*darne* (Fr), darne

*darne de saumon* (Fr), darne de sau-mon

*dartois* (Fr), dar-tois

*dasheen* (US), da-sheen

*dashi* (Jp), da-shi

*dà-syār* (Ch), dà-syār

*date* (US), date

*datle* (Cz), dat-le

*datte* (Fr), datte

*Datteln* (Gr), Dat-teln

*dattero* (It), dat-te-ro

*dattero di mare* (It), dat-te-ro di ma-re

*daube* (Fr), daube

*daube de moreton* (Fr), daube de mo-re-ton

*daumont, à la* (Fr), dau-mont, à la

*daun selada* (In), daun sela-da

*daun sup* (In), daun sup

*dauphine, à la* (Fr), dau-phine, à la

*daurade* (Fr), dau-rade

*debreceni rántotta* (Hu), deb-re-ce-ni rán-tot-ta

*deep-fry* (US), deep-fry

*deglaze* (US), de-glaze

*del giorno* (It), del gi-or-no

*délices* (Fr), dé-li-ces

*Delmonico* (US), Del-mon-i-co

*demerara sugar* (GB), deme-e-rar-a sug-ar

*demi-deuil* (Fr), de-mi-deu-il

*demi-glace* (Fr), dem-i-glace

*demi-tasse* (Fr), dem-i-tasse

*demoiselles de Maine* (Fr), de-moi-selles de Maine

*dendê* (Pg), dendê

*dendeng pedas* (In), den-deng pe-das

*dendeng ragi* (In), den-deng ra-gi

*dengaku* (Jp), den-ga-ku

*dent-de-lion* (Fr), dent-de-li-on

*dentelle* (Fr), den-telle

*Derby* (GB), Der-by

*derma* (US), der-ma

*désossé* (Fr), dé-sos-sé

*deviled* (US), dev-iled

*deviled ham* (US), dev-iled ham

*dextrose* (US), dex-trose

*dhania* (Ia), dhan-i-a

*dhansak* (Ia), dhan-sak

*dhoka* (Ia), dho-ka

*dhuli urad* (Ia), dhu-li u-rad

*diable, à la* (Fr), di-a-ble, à la

**diable, sauce à la** (Fr), di-a-ble, sauce à la
**diabloki** (Po), di-a-blo-ki
**diablotins de fromage** (Fr), di-a-blo-tins de fro-mage
**dia rau song** (Vt), dia rau song
**diavolo, alla** (It), di-a-vo-lo, al-la
**dibakar** (In), di-ba-kar
**dibs** (Ar), dibs
**dice** (US), dice
**Dicke Bohnen** (Gr), Dic-ke Boh-nen
**diente de ajo** (Sp), di-en-te de a-jo
**dieppoise, à la** (Fr), diep-poise, à la
**dietary fiber** (US), di-et-ar-y fi-ber
**digestif** (Fr), di-ges-tif
**digestive biscuit** (GB), di-ges-tive bis-cuit
**digestivi** (It), di-ges-ti-vi
**dijonnaise, sauce** (Fr), di-jon-naise, sauce
**dil** (Tr), dil
**dil baliği** (Tr), dil ba-liği
**dill** (US), dill
**dillkött** (Sw), dill-kött
**dilly beans** (US), dil-ly beans
**di mare** (It), di ma-re
**dim sum** (Ch), dim sum
**dinde** (Fr), dinde
**dinde rôtie** (Fr), dinde rô-tie
**dindonneau** (Fr), din-don-neau
**ding syang** (Ch), ding syang
**dinja** (SC), di-nja
**dinnye** (Hu), din-nye

**diós kifli** (Hu), di-ós kif-li
**diós metélt** (Hu), di-ós me-télt
**diples** (Gk), di-ples
**diplomat** (Fr), di-plo-mat
**dirty rice** (US), dirt-y rice
**disznóbús** (Hu), disz-nó-hús
**disznókocsonya** (Hu), disz-nó-ko-csonya
**ditali** (It), di-ta-li
**divinity** (US), di-vin-i-ty
**divljač** (SC), div-ljač
**djuveč** (SC), dju-več
**dobin mushi** (Jp), do-bin mu-shi
**dobostorta** (Hu), do-bos-tor-ta
**dobrada** (Pg), do-bra-da
**doce** (Pg), do-ce
**dodine de canard** (Fr), do-dine de ca-nard
**doen jaŋg** (Kr), doen jang
**dogfish** (US), dog-fish
**dojrzale oliwki** (Po), do-jr-za-le o-liw-ki
**dolce** (It), dol-ce
**dolce antico** (It), dol-ce an-ti-co
**dolce e agro** (It), dol-ce e a-gro
**dolce Maria** (It), dol-ce Ma-ri-a
**dolci** (It), dol-ci
**dolma** (Gk, Tr), dol-ma
**dolmadakia** (Gk), dol-ma-da-kia
**dolmades** (Gk), dol-ma-des
**domashniaia lapsha** (Rs), do-mash-nia-ia lap-sha
**domates** (Tr), do-ma-tes

*domatesli fasulye* (Tr), do-ma-tes-li fa-sul-ye

*domates yemistes me rizi* (Gk), do-ma-tes ye-mi-stes me riz-i

*domatorizo pilafi* (Gk), do-ma-to-riz-o pi-la-fi

*domba* (In), dom-ba

*domburi* (Jp), dom-bu-ri

*dôme blanche* (Fr), dôme blanche

*dom ke kai* (Th), dom ke kai

*domuz* (Tr), do-muz

*domyoji ko* (Jp), dom-yo-ji ko

*dondurma* (Tr), don-dur-ma

*doner* (Gk), do-ner

*döner kebabi* (Tr), dö-ner ke-ba-bi

*doopiaza* (Ia), doo-pi-a-za

*doree* (Fr), do-ree

*doreshingu* (Jp), do-re-shin-gu

*doriani* (Af-Swahili), do-ri-a-ni

*Dorsch* (Gr), Dorsch

*dort* (Cz), dort

*dosas* (Ia), do-sas

*dot, to* (US), dot, to

*dòu* (Ch), dòu

*double consommé* (US), dou-ble con-som-mé

*double cream* (GB), dou-ble cream

*double cream cheese* (US), dou-ble cream cheese

*dòu-fú* (Ch), dòu-fú

*dòu-fú chīng-tsài tāng* (Ch), dòu-fú chīng-tsài tāng

*dough* (US), dough

*doughboy* (US), dough-boy

*doughnut* (US), dough-nut

*doughnut hole* (US), dough-nut hole

*dòu-shā bāu* (Ch), dòu-shā bāu

*doux* (Fr), doux

*dòu-yá-tsài* (Ch), dòu-yá-tsài

*Dover sole* (GB), Do-ver sole

*dragées* (Fr), dra-gées

*draw, to* (US), draw, to

*drawn butter* (US), drawn but-ter

*dredge, to* (US), dredge, to

*dress, to* (US), dress, to

*dressing* (US), dress-ing

*dried meat* (US), dried meat

*drizzle* (US), driz-zle

*drop biscuit* (US), drop bis-cuit

*drozhzhevoe testo* (Rs), drozh-zhe-voe tes-to

*dršťková polévka* (Cz), dršť-ko-vá po-lév-ka

*druer* (Nw), dru-er

*druiven* (Du), drui-ven

*drum* (US), drum

*drupe fruit* (US), drupe fruit

*dry curries* (Ia), dry cur-ries

*dry milk* (US), dry milk

*du Barry, à la* (Fr), du Bar-ry, à la

*Dublin coddle* (Ir), Dub-lin cod-dle

*duchesse* (Fr), duch-esse

*duchesse potatoes* (US), duch-esse po-ta-toes

*duck* (US), duck

*duck ham* (US), duck ham

*duck sauce* (US), duck sauce

*due* (Nw), du-e

*duff* (GB), duff
*duglère, à la* (Fr), du-glère, à la
*dulce* (Sp), dul-ce
*dulce de elote* (Sp), dul-ce de e-lo-te
*dulce de higos* (Sp), dul-ce de hi-gos
*dulse* (US), dulse
*dum* (Ia), dum
*dum alu* (Ia), dum a-lu
*dumplings* (US), dump-lings
*Dundee cake* (Sc), Dun-dee cake
*Dungeness crab* (US), Dun-ge-ness crab
*dūng-gŭ* (Ch), dūng-gŭ
*dūng-gwā* (Ch), dūng-gwā
*Dunlop* (Sc), Dun-lop
*durazno* (Sp), du-raz-no
*durazno en crema* (Sp), du-raz-no en cre-ma
*durian* (In), du-ri-an

*durum wheat* (US), du-rum wheat
*duru-ten* (Jp), du-ru-ten
*duruwakashii* (Jp), du-ru-wa-ka-shi-i
*Duse, à la* (Fr), Duse, à la
*dušené* (Cz), du-še-né
*dust, to* (US), dust, to
*duva* (Sw), du-va
*duxelles* (Fr), dux-elles
*dyăn-syīn* (Ch), dyăn-syīn
*dyàu-wèi-pin* (Ch), dyàu-wèi-pin
*dybbavsreje* (Da), dyb-havs-re-je
*dynia* (Po), dy-nia
*dyrestek* (Nw), dyre-stek
*dzău-dz* (Ch), dzău-dz
*džem* (SC, Cz), džem
*dzieczyzna* (Po), dzie-czyz-na
*džigerica* (SC), dži-ge-ri-ca
*dzsem* (Hu), dzsem

# ℰ

*é* (Ch), é
*e* (It), e
*ears* (US), ears
*earshell* (US), ear-shell
*eau* (Fr), eau
*eau-de-vie* (Fr), eau-de-vie

*eau-de-vie de poire* (Fr), eau-de-vie de poire
*ebi* (In), e-bi
*ebi* (Jp), e-bi
*ebi furai* (Jp), e-bi fu-ra-i
*ebi-sembei* (Jp), ebi-sem-bei

*ebi suimono* (Jp), e-bi su-i-
mo-no

*écarlate, á l'* (Fr), é-car-late,
á l'

*Ecclefechan tart* (Sc), Ec-cle-
fech-an tart

*Eccles cake* (GB), Ec-cles cake

*ecet* (Hu), e-cet

*échalote* (Fr), é-cha-lote

*echaude* (Fr), e-chaude

*éclade* (Fr), é-clade

*éclair* (Fr), é-clair

*écossaise, à l'* (Fr), é-cos-
saise, à l'

*écrevisse* (Fr), é-cre-visse

*écrevisses à la nage* (Fr),
é-cre-vis-ses à la nage

*Edam* (US), E-dam

*edamame* (Jp), e-da-ma-me

*Edammer kaas* (Du), E-dam-
mer kaas

*eddik* (Nw), ed-dik

*eddike* (Da), ed-di-ke

*Edelkastanie* (Gr), E-del-
kas-tan-ie

*édeskömény* (Hu), é-des-kö-
mény

*édesnemes* (Hu), é-des-ne-
mes

*édességet* (Hu), é-des-sé-get

*Edinburgh fog* (Sc), Ed-in-
burgh fog

*EDTA* (US), EDTA

*eel* (US), eel

*eend* (Du), eend

*efterrätt* (Sw), ef-ter-rätt

*egg* (US), egg

*egg bløtkokt* (Nw), egg bløt-
kokt

*egg bread* (US), egg bread

*egg butter* (US), egg but-ter

*egg cream* (US), egg cream

*eggerøre* (Nw), egg-e-rø-re

*egg flip* (GB), egg flip

*egg foo yung* (US), egg foo
yung

*egg forlorne* (Nw), egg for-
lor-ne

*egg hårdkokt* (Nw), egg
hård-kokt

*eggnog* (US), egg-nog

*egg og bacon* (Nw), egg og
ba-con

*egg pie* (GB), egg pie

*eggplant* (US), egg-plant

*eggplant caviar* (US), egg-
plant cav-i-ar

*egg roll* (US), egg roll

*eggs Benedict* (US), eggs
Ben-e-dict

*eggs Sardou* (US), eggs Sardou

*egg wash* (US), egg wash

*églefin* (Fr), é-gle-fin

*egres* (Hu), e-gres

*ehu* (Pl), e-hu

*Eier* (Gr), Ei-er

*eieren* (Du), ei-e-ren

*eieren met ham* (Du), ei-e-
ren met ham

*eieren met spek* (Du), ei-e-
ren met spek

*eierenpannekoeken* (Du),
ei-e-ren-pan-ne-koe-ken

*Eierfrüchte* (Gr), Ei-er-früch-
te

*eiergebak* (Du), ei-er-ge-hak

***Eierkrem*** (Gr), Ei-er-krem
***Eierkuchen*** (Gr), Ei-er-ku-chen
***Eier mit Speck*** (Gr), Ei-er mit Speck
***Eiersalat*** (Gr), Ei-er-sa-lat
***Eierschaum*** (Gr), Ei-er-schaum
***Eierspeisen*** (Gr), Ei-er-spei-sen
***Einbrenn*** (Gr), Ein-brenn
***Eingemachte*** (Gr), Ein-ge-mach-te
***eiró*** (Pg), ei-ró
***Eis*** (Gr), Eis
***Eisen*** (Gr), Ei-sen
***Eisenbein*** (Gr), Ei-sen-bein
***Eiskrem*** (Gr), Eis-krem
***Eistee*** (Gr), Eis-tee
***Eistorte*** (Gr), Eis-tor-te
***Eiswein*** (Gr), Eis-wein
***ejotes*** (Mx), e-jo-tes
***ejotes con limon*** (Mx), e-jo-tes con li-mon
***ekmek*** (Tr), ek-mek
***ekmek kataifi*** (Gk), ek-mek ka-ta-i-fi
***ekşi*** (Tr), ek-şi
***elaichi*** (Ia), e-lai-chi
***elaichi murgh*** (Ia), e-lai-chi murgh
***Elbo*** (Da), El-bo
***elbow pastas*** (US), el-bow pas-tas
***elderberry*** (US), el-der-ber-ry
***election cake*** (US), e-lec-tion cake
***Elefantenlaus*** (Gr), E-le-fan-ten-laus

***eleoladho*** (Gk), e-leo-la-dho
***elft*** (Du), elft
***elies*** (Gk), e-lies
***ellinikous mezedhes*** (Gk), el-li-ni-kous mez-e-dhes
***elma*** (Tr), el-ma
***elote asado*** (Mx), e-lo-te a-sa-do
***elvers*** (US), el-vers
***embe*** (Af-Swahili), e-mbe
***émincés*** (Fr), é-min-cés
***Emmentaler*** (Gr-Switzer-land), Em-men-tal-er
***empada*** (Pg), em-pa-da
***empada de galinha*** (Pg), em-pa-da de ga-li-nha
***empal*** (In), em-pal
***empalpedas*** (In), em-pal-pe-das
***empanada*** (Sp), em-pa-na-da
***empanada de pulpo*** (Sp), em-pa-na-da de pul-po
***empanada de verde*** (Sp), em-pa-na-da de ver-de
***empanaditas*** (Sp), em-pa-na-di-tas
***empanaditas de queso*** (Sp), em-pa-na-di-tas de que-so
***empanado*** (Sp), em-pa-na-do
***emulsifier*** (US), e-mul-si-fi-er
***emulsion*** (US), e-mul-sion
***en bianco*** (It), en bian-co
***en bordure*** (Fr), en bor-dure
***enchilada*** (Sp), en-chi-la-da
***en cocotte*** (Fr), en co-cotte
***en croûte*** (Fr), en croûte
***encurtidos*** (Sp), en-cur-ti-dos

*endive* (US), en-dive

*endive à la normande* (Fr), en-dive à la nor-mande

*endive belge* (Fr), en-dive belge

*endōmame* (Jp), en-dō-ma-me

*eneldo* (Sp), e-nel-do

*enginar* (Tr), en-gi-nar

*Englischer Kuchen* (Gr), Eng-li-scher Ku-chen

*English muffin* (US), Eng-lish muf-fin

*English walnut* (US), Eng-lish wal-nut

*enguia* (Pg), en-gui-a

*en meurette* (Fr), en meu-rette

*enoki* (Jp), e-no-ki

*enriched products* (US), en-riched prod-ucts

*enrollados* (Mx), en-ro-lla-dos

*ensalada* (Sp), en-sa-la-da

*ensalada variada* (Sp), en-sa-la-da va-ri-a-da

*Ente* (Gr), En-te

*entrecosto de porco* (Pg), en-tre-cos-to de por-co

*entrecôte* (Fr), en-tre-côte

*entrecôte à la Bordelaise* (Fr), en-tre-côte à la Bor-de-laise

*entrecôte aux cèpes* (Fr), en-tre-côte aux cèpes

*entrecôte marchand de vin* (Fr), en-tre-côte mar-chand de vin

*entrée* (Fr), en-trée

*entremets* (Fr), en-tre-mets

*épaule* (Fr), é-paule

*épaule d'agneau* (Fr), é-paule d'agneau

*épaule de veau* (Fr), é-paule de veau

*epazote* (Mx), e-pa-zo-te

*éperlan* (Fr), é-per-lan

*épice* (Fr), é-pice

*épigramme* (Fr), é-pi-gramme

*épinard* (Fr), é-pi-nard

*épinards au beurre noisette* (Fr), é-pi-nards au beurre noi-sette

*eple* (Nw), ep-le

*eplekake* (Nw), ep-le-kak-e

*equatorial cuisine* (US), e-qua-tor-i-al cui-sine

*érable* (Fr), é-ra-ble

*eramaahiekhaa* (Fi), e-ra-maa-hi-ek-haa

*erbe* (It), er-be

*Erbsen* (Gr), Erb-sen

*Erbsensuppe mit saurer Sahne* (Gr), Erb-sen-sup-pe mit sau-rer Sah-ne

*Erdäpfel* (Gr-Austria), Erd-äp-fel

*Erdartischocke* (Gr), Erd-ar-ti-scho-cke

*Erdbeeren* (Gr), Erd-bee-ren

*erik* (Tr), e-rik

*eros* (Hu), e-ros

*ersatz food* (US), er-satz food

*erter* (Nw), ert-er

*ervilhas* (Pg), er-vi-lhas

*erwtensoep* (Du), erw-ten-soep

*erwtjes* (Du), erw-tjes

*Esau* (Fr), E-sau

**Esbare Muscheln** (Gr), Es-ba-re Mu-scheln

**escabeche** (Sp), es-ca-be-che

**escalibada** (Sp), es-ca-li-ba-da

**escalopes** (Fr), es-cal-opes

**escalopes de bar à l'oscille** (Fr), es-cal-opes de bar à l'os-cille

**escalopes de sanglier** (Fr), es-cal-opes de sang-lier

**escalopes de saumon á l'oiseille** (Fr), es-cal-opes de sau-mon á l'oi-seille

**escalopes de ternera** (Sp), es-ca-lo-pes de ter-ne-ra

**escalopes de veau** (Fr), es-cal-opes de veau

**escalopes de veau cordon bleu** (Fr), es-cal-opes de veau cor-don bleu

**escalopes de veau sautées á l'estragon** (Fr), es-cal-opes de veau sau-tées á l'es-tra-gon

**escargots** (Fr), es-car-gots

**escargots à la bourguignonne** (Fr), es-car-gots à la bour-gui-gnonne

**escargots de Bourgogne** (Fr), es-car-gots de Bour-go-gne

**escarola alla monachina** (It), e-sca-ro-la al-la mon-a-chi-na

**escarole** (US), es-ca-role

**espadarte** (Pg), es-pa-dar-te

**espadilha** (Pg), es-pa-di-lha

**espadin** (Sp), es-pa-din

**espadon** (Fr), es-pa-don

**espagnole, à l'** (Fr), es-pa-gnole, à l'

**espagnole, sauce** (Fr), es-pa-gnole, sauce

**espargos** (Pg), es-par-gos

**espárragos** (Sp), es-pár-ra-gos

**espárragos con tomatillos** (Mx), e-spár-ra-gos con to-ma-ti-llos

**especialidade da casa** (Pg), es-pe-ci-a-li-da-de da ca-sa

**espinaca** (Sp), es-pi-na-ca

**espinacas a la Catalana** (Sp), es-pi-na-cas a la Ca-ta-la-na

**espinafre** (Pg), es-pi-na-fre

**espresso** (It), es-pres-so

**esprot** (Fr), es-prot

**essence** (US), es-sence

**essential amino acids** (US), es-sen-tial a-mi-no ac-ids

**essential fatty acids** (US), es-sen-tial fat-ty ac-ids

**essential nutrients** (US), es-sen-tial nu-tri-ents

**essential oils** (US), es-sen-tial oils

**Essig** (Gr), E-ssig

**Essigsäure** (Gr), E-ssig-säu-re

**Esterbazy rostbraten** (Gr), Es-ter-ha-zy rost-bra-ten

**estofado** (Sp), es-to-fa-do

**estofado almendrado de pollo** (Sp), es-to-fa-do al-men-dra-do de po-llo

**estofado de carnero** (Sp), es-to-fa-do de car-ne-ro

**estouffade** (Fr), es-touf-fade

*estouffade de boeuf* (Fr), es-touf-fade de boeuf
*estragon* (Fr), es-tra-gon
*esturgeon* (Fr), es-tur-geon
*esturión* (Sp), es-tu-ri-ón
*esturjão* (Pg), es-tur-jã-o
*et* (Tr), et
*etikka* (Fi), e-tik-ka
*étoile de beurre* (Fr), é-toile de beurre
*étouffée* (Fr), é-touf-fée

*et suyu* (Tr), et su-yu
*étuvé* (Fr), é-tu-vé
*étuver* (Fr), é-tu-ver
*evaporated milk* (US), e-vap-o-rat-ed milk
*ewe* (US), ewe
*exohiko* (Gk), ex-o-hi-ko
*extracts* (US), ex-tracts
*extra grouse* (Aa), ex-tra grouse
*extra-sec* (Fr), ex-tra-sec

# F

*fabada* (Sp), fa-ba-da
*fadge* (Ir), fadge
*faggot* (GB), fag-got
*faggots* (GB), fag-gots
*fagiano* (It), fa-gia-no
*fagioli* (It), fa-gio-li
*fagioli al fiasco* (It), fa-gio-li al fi-as-co
*fagioli assoluti* (It), fa-gio-li as-so-lu-ti
*fagioli con le cotiche* (It), fa-gio-li con le co-ti-che
*fagioli di lima* (It), fa-gio-li di li-ma
*fagiolini* (It), fa-gio-li-ni
*fagiolini in padella* (It), fa-gio-li-ni in pa-del-la
*fagioli toscani col tonno* (It), fa-gio-li tos-ca-ni col ton-no

*fagottini* (It), fa-got-ti-ni
*fagottini di melanzane* (It), fa-got-ti-ni di me-lan-za-ne
*fagyasztott* (Hu), fa-gyasz-tott
*faisan* (Fr), fai-san
*faisán* (Sp), fai-sán
*faizan s gribami v smetane* (Rs), fai-zan s grib-a-mi v sme-ta-ne
*fajita* (Sp), fa-ji-ta
*fakha* (Ar), fak-ha
*faki* (Gk), fa-ki
*falafel* (Ar), fa-la-fel
*falsa* (Ia), fal-sa
*falukorv* (Sw), fa-lu-korv
*fàn* (Ch), fàn
*fanchonette* (Fr), fan-chon-ette
*fān-chyé* (Ch), fān-chyé

*fān-chyé shā-là* (Ch), fān-chyé shā-là
*fān-chyé tāng* (Ch), fān-chyé tāng
*farala* (Ar), fa-ra-la
*faraona* (It), fa-rao-na
*faraona al cartoccio* (It), fa-rao-na al car-toc-cio
*farce de poisson* (Fr), farce de pois-son
*farcement* (Fr), farce-ment
*farces* (Fr), far-ces
*farci* (Fr), far-ci
*farcito* (It), far-ci-to
*fårekjøtt* (Nw), få-re-kjøtt
*fårekotelett* (Nw), få-re-ko-te-lett
*fårerull* (Nw), få-re-rull
*farfale* (It), far-fa-le
*farfel* (Jw), far-fel
*får i kål* (Nw), får i kål
*farin* (Nw), fa-rin
*farina* (US), fa-ri-na
*farina dolce* (It), fa-ri-na dol-ce
*farina gialla* (It), fa-ri-na gi-al-la
*farine* (Fr), fa-rine
*farine de maïs* (Fr), fa-rine de maïs
*farinha* (Pg), fa-ri-nha
*farinha de aveia* (Pg), fa-ri-nha de a-ve-ia
*farkha* (Ar-Egypt), far-kha
*fårkött* (Sw), får-kött
*farl* (Sc), farl
*farmer's cheese* (US), farmer's cheese
*farofa* (Pg-Brazil), fa-ro-fa
*farófias* (Pg), fa-ró-fi-as

*farro* (Sp), far-ro
*farshirovanniye kambala* (Rs), far-shi-ro-van-ni-ye kam-ba-la
*farshirovanniye luk* (Rs), far-shi-ro-van-ni-ye luk
*fårsk oxbringa* (Sw), fårsk ox-bring-a
*färskt* (Sw), färskt
*fasan* (Da, Nw), fa-san
*Fasan* (Gr), Fa-san
*fasaner* (Da), fa-sa-ner
*Fasan mit Weinkraut* (Gr), Fa-san mit Wein-kraut
*Fasan mit weissen Bohnen* (Gr), Fa-san mit weis-sen Boh-nen
*Faschierter Braten* (Gr), Fasch-ier-ter Bra-ten
*faséole* (Fr), fas-é-ole
*faširane šnicle* (SC), fa-ši-ra-ne šni-cle
*fasirozott* (Hu), fa-si-ro-zott
*fasola* (Po), fa-so-la
*fasolia* (Gk), fa-so-lia
*fasoulakia* (Gk), fa-sou-la-kia
*fast food* (US), fast food
*Fastnachtkrapfen* (Gr), Fast-nacht-krap-fen
*fasul* (Bu), fa-sul
*fasulye* (Tr), fa-sul-ye
*fasulyeli paça* (Tr), fa-sul-ye-li pa-ça
*fasulye pilâkisi* (Tr), fa-sul-ye pi-lâ-kisi
*fatányéros* (Hu), fa-tá-nyé-ros
*fatayer* (Ar), fa-ta-yer
*fatback* (US), fat-back

**fat-free** (US), fat-free
**fatia** (Pg), fa-ti-a
**fatias frias** (Pg), fa-ti-as fri-as
**fatir** (Ar), fa-tir
**fats and oils** (US), fats and oils
**fatta** (Ar), fat-ta
**fattiga riddare** (Sw), fat-ti-ga rid-da-re
**fatto in casa** (It), fat-to in ca-sa
**fattoush** (Ar), fat-toush
**fatty acids** (US), fat-ty ac-ids
**faux mousseron** (Fr), faux mous-se-ron
**faux-nuts** (US), faux-nuts
**fava bean** (It), fa-va bean
**fava frescas em salada** (Pg), fa-va fres-cas em sa-la-da
**Fayyoum duck** (Ar), Fay-youm duck
**fazan po-Gruzinski** (Rs), fa-zan po-Gru-zin-ski
**fazant** (Du), fa-zant
**fazole** (Cz), fa-zo-le
**fazolky** (Cz), fa-zol-ky
**fazolový salát** (Cz), fa-zo-lo-vý sa-lát
**fegatini di pollo alla salvia** (It), fe-ga-ti-ni di pol-lo al-la sal-vi-a
**fegato** (It), fe-ga-to
**fegato alla veneziana** (It), fe-ga-to al-la ve-ne-zi-a-na
**fegato di pollo alla salvia** (It), fe-ga-to di pol-lo al-la sal-vi-a
**fegato di vitello** (It), fe-ga-to di vi-tel-lo
**febérbort** (Hu), fe-hér-bort

**febér kenyér** (Hu), fe-hér ken-yér
**Feigen** (Gr), Fei-gen
**feijão** (Pg), fei-jão
**feijão guisado** (Pg), fei-jão gui-sa-do
**feijão de vagens** (Pg), fei-jão de va-gens
**feijoada** (Pg), fei-joa-da
**Feingebäck** (Gr), Fein-ge-bäck
**fejessaláta** (Hu), fe-jes-sa-lá-ta
**femöring med ägg** (Sw), fe-mö-ring med ägg
**fenalår** (Nw), fen-a-lår
**Fenchel** (Gr), Fen-chel
**fenesi** (Af-Swahili), fe-ne-si
**feng wei hsia** (Ch), feng wei hsia
**fen kuo** (Ch), fen kuo
**fennel seeds** (US), fen-nel seeds
**fenouil** (Fr), fen-ouil
**fenouil, au** (Fr), fen-ouil, au
**fenugreek** (US), fen-u-greek
**fer** (Fr), fer
**féra** (Fr), fé-ra
**ferique** (Ar), fe-rique
**fermière, à la** (Fr), fer-mière, à la
**ferri, ai** (It), fer-ri, ai
**fersken** (Da, Nw), fers-ken
**fesa di vitello** (It), fe-sa di vi-tel-lo
**fesanjan** (Ar), fes-an-jan
**Feta** (Gk), Fe-ta
**fette** (It), fet-te
**fettine di manzo alla piz-**

**zaiola** (It), fet-ti-ne di man-zo al-la piz-za-io-la

**fettucine** (It), fet-tu-ci-ne

**fettucine al burro** (It), fet-tu-ci-ne al bur-ro

**fettucine alla panna** (US), fet-tu-ci-ne al-la pan-na

**fettucine alla papalina** (It), fet-tu-ci-ne al-la pa-pa-li-na

**fettucine alla romano** (It), fet-tu-ci-ne al-la ro-man-o

**fettucine all'uova** (It), fet-tu-ci-ne all'uo-va

**feuilles de betterave** (Fr), feu-illes de bet-te-rave

**Feuilles de Dreux** (Fr), Feu-illes de Dreux

**feuilles de vigne** (Fr), feu-il-les de vi-gne

**feuilleté** (Fr), feu-ille-té

**feuilleté au fromage** (Fr), feu-ille-té au fro-mage

**feuilleté de fruits de mer** (Fr), feu-ille-té de fruits de mer

**feuilleté de homard** (Fr), feu-ille-té de ho-mard

**feuilleté de ris de veau** (Fr), feu-ille-té de ris de veau

**fèves** (Fr), fèves

**fèves au lard** (Fr-Canada), fèves au lard

**fèves de marais** (Fr), fèves de ma-rais

**fiambre** (Pg), fi-am-bre

**fiambres** (Sp), fi-am-bres

**fiber** (US), fi-ber

**ficat de pui cu ceapă** (Ro), fi-cat de pui cu ce-a-pă

**fichi** (It), fi-chi

**fiddleheads** (US), fid-dle-heads

**fideos gordos** (Sp), fi-de-os gor-dos

**fidget pie** (GB), fidg-et pie

**fig** (US), fig

**figado** (Pg), fi-ga-do

**figili** (Af-Swahili), fi-gi-li

**figl** (Ar), figl

**figos** (Pg), fi-gos

**figues** (Fr), figues

**fiken** (Nw), fi-ken

**fikon** (Sw), fi-kon

**fiky** (Cz), fi-ky

**filbert** (US), fil-bert

**filé** (US), fi-lé

**filet** (US), fi-let

**filet, to** (US), fil-et, to

**filet de boeuf** (Fr), fi-let de boeuf

**filet de boeuf en croûte** (Fr), fi-let de boeuf en croûte

**filet de sole amandine** (Fr), fi-let de sole a-man-dine

**filete de cerdo** (Sp), fi-le-te de cer-do

**filet mignon** (Fr), fi-let mi-gnon

**filets d'anchois** (Fr), fi-lets d'an-chois

**filets de hareng de la baltique crème** (Fr), fi-lets de ha-reng de la bal-tique crème

**filetto di bue** (It), fi-let-to di bu-e

**filetto di manzo** (It), fi-let-to di man-zo

**filetto di pesce oreganato**

(It), fi-let-to di pes-ce o-re-ga-na-to

*filetto di sogliole* (It), fi-let-to di sog-lio-le

*filetto di tacchino alla crema* (It), fi-let-to di ta-cchi-no al-la cre-ma

*filfil ahkdar* (Ar), fil-fil ahk-dar

*filfil mahchi* (Ar), fil-fil mah-chi

*filbó* (Pg), fi-lhó

*fillet* (US), fil-let

*fillet, to* (US), fil-let, to

*filmjölk* (Sw), film-jölk

*filo* (Gk), fi-lo

*filosoof* (Du), fil-o-soof

*financière, à la* (Fr), fi-nan-cière, à la

*finanziera di pollo* (It), fi-nan-zi-er-a di pol-lo

*finbrød* (Nw), fin-brød

*findik* (Tr), fin-dik

*fines herbes* (Fr), fines herbes

*finger pies* (US), fin-ger pies

*finnan haddie* (Sc), fin-nan had-die

*finocchi* (It), fi-noc-chi

*finocchio al forno* (It), fi-noc-chi-o al for-no

*finsiktat mjol* (Sw), fin-sik-tat mjol

*finta* (Rs), fin-ta

*fiorentina, alla* (It), fio-ren-ti-na, al-la

*fiori di zucchini* (It), fio-ri di zuc-chi-ni

*fiori fritti* (It), fio-ri frit-ti

*fi qa'atah* (Ar), fi qa'a-tah

*fireek* (Ar), fi-reek

*fire pot* (US), fire pot

*firinda* (Tr), fi-rin-da

*firnee* (Ia), fir-nee

*Fisch* (Gr), Fisch

*Fischbrühe* (Gr), Fisch-brü-he

*fischietti* (It), fisch-iet-ti

*Fischmayonnaise* (Gr), Fisch-may-on-nai-se

*Fischrouladen* (Gr), Fisch-rou-la-den

*Fischschüssel* (Gr), Fisch-schüs-sel

*fish slice* (GB), fish slice

*fish sticks* (US), fish sticks

*fish tartar* (Jw), fish tar-tar

*fisk* (Da, Nw), fisk

*fiskeboller* (Nw), fis-ke-boll-er

*fiskegratin* (Nw), fis-ke-gra-tin

*fiskekaker* (Nw), fis-ke-kak-er

*fiskepudding* (Nw), fis-ke-pud-ding

*fisksoppa* (Sw), fisk-sop-pa

*Fisolen* (Gr-Austria), Fi-so-len

*five-spice powder* (US), five-spice pow-der

*fjaerfe* (Nw), fjaer-fe

*fläderbär* (Sw), flä-der-bär

*Flädle* (Gr), Flä-dle

*Flädlesuppe* (Gr), Flä-dle-sup-pe

*flæskesteg* (Da), flæs-ke-steg

*flæskeæggekage* (Da), flæs-ke-æg-ge-ka-ge

*flaesk i kål* (Da), flaesk i kål

*flageolet* (Fr), fla-geo-let
*flake, to* (US), flake, to
*flaki po polsku* (Po), fla-ki po pol-sku
*flamande, à la* (Fr), fla-mande, à la
*flambé* (Fr), flam-bé
*flamiche* (Fr), fla-miche
*flan* (Sp), flan
*flan* (Fr), flan
*flan au lait* (Fr), flan au lait
*flanchet* (Fr), flan-chet
*flan de manzanas* (Sp), flan de man-za-nas
*flank steak* (US), flank steak
*flannel cake* (US), flan-nel cake
*flapjack* (US), flap-jack
*fläsk* (Sw), fläsk
*fläsk korv* (Sw), fläsk korv
*flatbrød* (Nw), flat-brød
*flatfish* (US), flat-fish
*flauta* (Mx), flau-ta
*Flecke* (Gr), Flec-ke
*Fleischbrühe* (Gr), Fleisch-brü-he
*Fleischklöschen* (Gr), Fleisch-klös-chen
*flensjes* (Du), flen-sjes
*flesk* (Nw), flesk
*flet* (Fr), flet
*flétan* (Fr), flé-tan
*fleurons* (Fr), fleur-ons
*fleurs de courgettes* (Fr), fleurs de cour-gettes
*flingor* (Sw), fling-or
*flitch* (GB), flitch
*floating island* (US), float-ing is-land

*flocos de cereal* (Pg), flo-cos de ce-re-al
*fløde* (Da), flø-de
*fløderand* (Da), flø-de-rand
*fløde skum* (Da), flø-de skum
*florentine, à la* (Fr), flor-en-tine, à la
*flory* (Sc), flo-ry
*fløte* (Nw), flø-te
*fløtevaffle* (Nw), flø-te-vaf-fle
*flounder* (US), floun-der
*flour* (US), flour
*fluke* (US), fluke
*flummery* (US), flum-mer-y
*flummery* (GB), flum-mer-y
*Flunder* (Gr), Flun-der
*flundra* (Sw), flun-dra
*fluoride* (US), fluor-ide
*Fluskrebs* (Gr), Flus-krebs
*flyingfish* (Cb), fly-ing-fish
*flyndre* (Nw), flyn-dre
*focaccia* (It), fo-cac-cia
*focaccia alla salvia* (It), fo-cac-cia al-la sal-vi-a
*focaccia di vitello* (It), fo-cac-cia di vi-tel-lo
*fogas* (Hu), fo-gas
*foie* (Fr), foie
*foie de canard* (Fr), foie de can-ard
*foie de poulet* (Fr), foie de pou-let
*foie de veau* (Fr), foie de veau
*foie de veau auvergnate* (Fr), foie de veau au-ver-gnate
*foie de volaille en brochette* (Fr), foie de vo-laille en bro-chette

*foie gras* (Fr), foie gras
*foie gras de canard en gelée* (Ca), foie gras de can-ard en ge-lée
*foie gras en croûte* (Fr), foie gras en croûte
*foie gras truffé* (Fr), foie gras truf-fé
*foiolo* (It), fo-io-lo
*fokhagyma* (Hu), fok-hagy-ma
*folate* (US), fo-late
*fold, to* (US), fold, to
*folic acid* (US), fo-lic ac-id
*folyami rák* (Hu), fo-lya-mi rák
*fond* (Fr), fond
*fondant* (Fr), fon-dant
*fondant* (US), fon-dant
*fondas de alcachofas* (Sp), fon-das de al-ca-cho-fas
*fond blanc* (Fr), fond blanc
*fond de volaille* (Fr), fond de vo-laille
*fonds d'artichaut* (Fr), fonds d'ar-ti-chaut
*fondue* (Fr), fon-due
*fondue bourguignonne* (Fr), fon-due bour-gui-gnonne
*fonduta* (It), fon-du-ta
*Fontainbleau* (Fr), Fon-tain-bleau
*fontina Val d'Aosta* (It), fon-ti-na Val d'A-os-ta
*foo foo* (Cb), foo foo
*fool* (GB), fool
*forcemeat* (US), force-meat
*forel* (Du), fo-rel
*forell* (Sw), fo-rell
*Forelle* (Gr), Fo-rel-le

*Forelle blau* (Gr), Fo-rel-le blau
*Forellen in Gurkensosse* (Gr), Fo-rel-len in Gurk-en-sos-se
*forel v vino* (Rs), fo-rel v vi-no
*forestière, à la* (Fr), for-es-tière, à la
*forfar bridies* (Sc), for-far brid-ies
*forloren* (Da), for-lo-ren
*formaggio* (It), for-mag-gio
*formaggio di crema* (It), for-mag-gio di cre-ma
*formato* (It), for-ma-to
*formato di carne* (It), for-ma-to di car-ne
*forno, al* (It), for-no, al
*forno, no* (Pg), for-no, no
*forretter* (Nw), for-ret-ter
*forró csokoládé* (Hu), for-ró cso-ko-lá-dé
*forshchmak* (Rs), forsh-chmak
*Förstertopf mit Pilzen* (Gr), För-ster-topf mit Pil-zen
*fortune cookie* (US), for-tune cook-ie
*fött krumpli* (Hu), fött krump-li
*four, au* (Fr), four, au
*fourrage à la crème d'orange* (Fr), four-rage à la crème d'or-ange
*fourrer* (Fr), four-rer
*fowl* (US), fowl
*Fra Diavolo* (It), Fra Dia-vo-lo
*fragole* (It), fra-go-le

*fragoline* (It), fra-go-li-ne

*fragoline di mare* (It), fra-go-li-ne di ma-re

*frais* (Fr), frais

*fraises* (Fr), frai-ses

*fraises aux liqueurs* (Fr), frai-ses aux li-queurs

*fraises chantilly* (Fr), frai-ses chan-til-ly

*fraises des bois* (Fr), frai-ses des bois

*fraises Romanof* (Fr), frai-ses Ro-man-of

*framboesas* (Pg), fram-bo-e-sas

*framboises* (Fr), fram-bois-es

*frambozen* (Du), fram-bo-zen

*frambuesas* (Sp), fram-bu-e-sas

*francala* (Tr), fran-ca-la

*Franconia potatoes* (US), Fran-con-ia po-ta-toes

*frangipane* (US), fran-gi-pane

*frango* (Pg), fran-go

*frango no espota à moda de Minho* (Pg), fran-go no es-po-ta à mo-da de Mi-nho

*frankfurters* (US), frank-furt-ers

*Frankfurter Würstchen* (Gr), Frank-fur-ter Würst-chen

*Frankfurterplatte* (Gr), Frank-fur-ter-plat-te

*frappé* (Fr), frap-pé

*freddo* (It), fred-do

*free-range* (US), free-range

*freestone* (US), free-stone

*freezer burn* (US), freez-er burn

*French bean* (US), French bean

*French cream* (US), French cream

*French cuisine* (US), French cui-sine

*French dressing* (US), French dress-ing

*French fries* (US), French fries

*French pastry* (US), French past-ry

*French toast* (US), French toast

*fresas* (Sp), fre-sas

*fresca* (It), fres-ca

*fresh* (US), fresh

*Fresno chili* (US), Fres-no chil-i

*friandises* (Fr), fri-an-di-ses

*fricadelles* (Fr), fri-ca-delles

*fricandeau* (Fr), fri-can-deau

*fricandel* (Du), fric-an-del

*fricando* (Sp), fri-can-do

*fricassee* (US), fric-as-see

*fricassée* (Fr), fric-as-sée

*fried green tomatoes* (US), fried green to-ma-toes

*fried rice* (US), fried rice

*fries* (GB), fries

*frigărui* (Ru), fri-gă-rui

*frijolada* (Sp), fri-jo-la-da

*frijoles* (Sp), fri-jo-les

*frijoles negros* (Sp), fri-jo-les ne-gros

*frijoles refritos* (Mx), fri-jo-les re-fri-tos

*Frikadellen* (Gr), Fri-ka-del-
len
*frikadeller* (Da, Sw), fri-ka-
del-ler
*frikase* (Gr), fri-ka-se
*frikassé* (Nw), fri-kas-sé
*Frikassee vom Huhn* (Gr),
Fri-kas-see vom Huhn
*frio* (Sp), fri-o
*frisch* (Gr), frisch
*friss* (Hu), friss
*frit* (Fr), frit
*frites* (Fr), frites
*frito* (Pg, Sp), fri-to
*fritots* (Fr), fri-tots
*frittata* (It), frit-ta-ta
*frittata con funghi* (It), frit-
ta-ta con fun-ghi
*frittata primavera* (It), frit-
ta-ta pri-ma-ve-ra
*fritte de ciliegielle* (It), fritte de ci-lie-giel-le
*frittelle* (It), frit-tel-le
*frittelle di polenta* (It), frit-
tel-le di po-len-ta
*fritter* (US), frit-ter
*fritto* (It), frit-to
*fritto misto* (It), frit-to mis-to
*fritto misto di mare* (It), frit-
to mis-to di ma-re
*fritura de pechugas de
pollo* (Sp), fri-tu-ra de pe-
chu-gas de po-llo
*fritura mixta* (Sp), fri-tu-ra
mix-ta
*friture* (Fr), fri-ture
*frizzle* (US), friz-zle
*frog's legs* (US), frog's legs
*froid* (Fr), froid
*fromage* (Fr), fro-mage

*fromage à la crème* (Fr), fro-
mage à la crème
*fromage blanc* (Fr), fro-
mage blanc
*fromage de tète* (Fr), fro-
mage de tète
*fromage du pays* (Fr), fro-
mage du pa-ys
*fromage rapé* (Fr), fro-mage
ra-pé
*Froschschenkel* (Gr), Frosch-
schen-kel
*frost, to* (US), frost, to
*frosted* (US), frost-ed
*frosting* (US), frost-ing
*Fruchtbrot* (Gr), Frucht-brot
*Fruchtkaltschale* (Gr),
Frucht-kalt-scha-le
*Fruchtpasteten* (Gr), Frucht-
pas-te-ten
*Fruchtsalat* (Gr), Frucht-sa-
lat
*fructose* (US), fruc-tose
*frugt* (Da), frugt
*frugtsaft* (Da), frugt-saft
*frugttærte* (Da), frugt-tær-te
*Frühlingskäse* (Gr-Austria),
Früh-lings-kä-se
*fruit* (Du), fruit
*fruit butters* (US), fruit butters
*fruitcake* (US), fruit-cake
*fruit cocktail* (US), fruit
cock-tail
*fruit cup* (GB), fruit cup
*fruit givré* (Fr), fruit giv-ré
*fruit salad* (US), fruit sal-ad
*fruits cuits* (Fr), fruits cuits
*fruits de mer* (Fr), fruits de
mer

*fruit sec* (Fr), fruit sec
*fruits frais* (Fr), fruits frais
*frukt* (Nw), frukt
*fruktkompott* (Nw), frukt-kom-pott
*fruktsaft* (Nw), frukt-saft
*frullato* (It), frul-la-to
*frumenta* (It), fru-men-ta
*frumenty* (GB), fru-men-ty
*fruset* (Sw), fru-set
*fruta* (Pg), fru-ta
*fruta azucarada* (Sp), fru-ta a-zu-ca-ra-da
*frutta cotta* (It), frut-ta cot-ta
*frutta della stagione* (It), frut-ta del-la sta-gio-ne
*frutta fresca* (It), frut-ta fres-ca
*frutta secca* (It), frut-ta sec-ca
*frutti di mare* (It), frut-ti di ma-re
*frutti misti* (It), frut-ti mis-ti
*fry* (US), fry
*fry, to* (US), fry, to
*fryer* (US), fry-er
*fu* (Jp), fu
*fudge* (US), fudge
*fugath* (Ia), fu-gath
*füge* (Hu), fü-ge
*fugl* (Nw), fugl
*fugu* (Jp), fu-gu
*fugusashi* (Jp), fu-gu-sa-shi
*fuki* (Jp), fu-ki
*fukujinzuke* (Jp), fu-ku-jin-zu-ke
*ful* (Ar), ful
*Füllung* (Gr), Fül-lung

*ful medamis* (Ar), ful me-da-mis
*fumé* (Fr), fu-mé
*fumet* (Fr), fu-met
*fundido* (Sp), fun-di-do
*funghetto, al* (It), fun-ghet-to, al
*funghi* (It), fun-ghi
*funghi acciugati al forno* (It), fun-ghi ac-ciu-ga-ti al for-no
*funghi porcini* (It), fun-ghi por-ci-ni
*funghi ripieni* (It), fun-ghi rip-ien-i
*funghi sott'olio ed aceto* (It), fun-ghi sot-t'ol-io ed a-ce-to
*funghi trifolati* (It), fun-ghi tri-fo-la-ti
*fun gwor* (Ch), fun gwor
*funnel cake* (US), fun-nel cake
*furaimono* (Jp), fu-rai-mo-no
*fursadi* (Af-Swahili), fur-sa-di
*fursecuri* (Ru), fur-se-cu-ri
*fú-rŭng dàn* (Ch), fú-rŭng dàn
*fusilli* (It), fu-sil-li
*füszeres* (Hu), fü-sze-res
*füszeres hozzávaló* (Hu), fü-sze-res hoz-zá-va-ló
*fyldt hvidkålshoved* (Da), fyldt hvid-kåls-ho-ved
*fylt kålbode* (Nw), fylt kål-ho-de
*Fynbo* (Da), Fyn-bo

# G

**gaai ka gosht** (Ia), gaai ka gosht

**gâche** (Fr), gâche

**gädda** (Sw), gäd-da

**gado-gado** (In), ga-do-ga-do

**gadon tahu** (In), ga-don tahu

**gaffelbitar** (Sw), gaf-fel-bi-tar

**Gaiskäsle** (Gr), Gais-käs-le

**gajar** (Ia), ga-jar

**gajar ka halva** (Ia), ga-jar ka hal-va

**galabart** (Fr), gal-ab-art

**galamb** (Hu), ga-lamb

**galangal** (US), ga-lan-gal

**galantine** (Fr), gal-an-tine

**galatoboureko** (Gk), ga-la-to-bou-re-ko

**galette** (Fr), ga-lette

**galettes aux fruits de mer** (Fr), ga-lettes aux fruits de mer

**galingale** (US), ga-lin-gale

**galinha** (Pg), ga-li-nha

**galinha assada** (Pg), ga-li-nha a-ssa-da

**gallimaufry** (GB), gal-li-mau-fry

**gallina a la pimienta** (Sp), gal-li-na a la pi-mi-en-ta

**gallo pinto** (Sp), gal-lo pin-to

**galushki** (Rs), ga-lush-ki

**galuska** (Hu), ga-lus-ka

**galuşte** (Ro), ga-luş-te

**gamba** (It), gam-ba

**gambari** (Ar), gam-ba-ri

**gambas** (Pg, Sp), gam-bas

**gambas a la plancha** (Sp, gam-bas a la plan-cha

**gamberello** (It), gam-be-rel-lo

**gamberetti di laguna** (It), gam-be-ret-ti di la-gu-na

**gamberetto grigio** (It), gam-be-ret-to gri-gio

**gamberi di fiume** (It), gam-be-ri di fiu-me

**game** (US), game

**game birds** (US), game birds

**game fish** (US), game fish

**Gammelost** (Nw), Gam-mel-ost

**gammon** (GB), gam-mon

**gamous** (Ar), ga-mous

**ganache** (Fr), ga-nache

**gandofli** (Ar), gan-dof-li

**ganmodoki** (Jp), gan-mo-do-ki

**gans** (Du), gans

**Gans** (Gr), Gans

*Gänseleberwurst* (Gr), Gän-se-le-ber-wurst
*ganso* (Pg), gan-so
*ganso relleno de castañas* (Sp), gan-so rel-le-no de cas-ta-ñas
*ga nuong ngu vi* (Vt), ga nu-ong ngu vi
*garam* (In), gar-am
*garam masala* (Ia), gar-am ma-sa-la
*garang asam ikan* (In), ga-rang a-sam i-kan
*garbanzo* (Sp), gar-ban-zo
*garbanzos salteados* (Sp), gar-ban-zos sal-te-a-dos
*garbure* (Fr), gar-bure
*garfish* (US), gar-fish
*garlic* (US), gar-lic
*garlic bread* (US), gar-lic bread
*garlic butter* (US), gar-lic but-ter
*garlic salt* (US), gar-lic salt
*garnalen* (Du), gar-na-len
*Garnele* (Gr), Gar-ne-le
*garni* (Fr), gar-ni
*garnish, to* (US), garn-ish, to
*garniture* (Fr), gar-ni-ture
*garretto* (It), gar-ret-to
*garulla* (Sp), ga-ru-lla
*garvie* (Sc), garv-ie
*gås* (Nw, Sw), gås
*gâscă prăjită* (Ro), găs-că pră-ji-tă
*gåsesteg* (Da), gå-se-steg
*gåsesteg med æbler og svedsker* (Da), gå-se-steg med æb-ler og sved-sker
*gastrique* (Fr), gas-trique

*gâteau* (Fr), gâ-teau
*gâteau à la brioche* (Fr), gâ-teau à la bri-oche
*gâteau de crêpes à la florentine* (Fr), gâ-teau de crêpes à la flor-en-tine
*gâteau de crêpes à la Normande* (Fr), gâ-teau de crêpes à la Nor-mande
*gatto* (It), ga-tto
*gaudes* (Fr), gau-des
*gauffres* (Bl), gauf-fres
*gaufrette* (Fr), gau-frette
*gauloise, à la* (Fr), gau-loise, à la
*gayette* (Fr), gay-ette
*gazar* (Ar), ga-zar
*gazpacho* (Sp), gaz-pa-cho
*Gebäck* (Gr), Ge-bäck
*gebacken* (Gr), ge-ba-cken
*gebakken* (Du), ge-bak-ken
*gebakken paling* (Du), ge-bak-ken pa-ling
*gebakken zeetong* (Du), ge-bak-ken zee-tong
*gebraden* (Du), ge-bra-den
*gebraden kip* (Du), ge-bra-den kip
*gebraten* (Gr), ge-bra-ten
*gedämft* (Gr), ge-dämft
*Geflügelsalat* (Gr), Ge-flü-gel-sa-lat
*gefillte fish* (Jw), ge-fill-te fish
*gefüllt* (Gr), ge-füllt
*gefüllte Kartoffeln* (Gr), ge-füll-te Kar-tof-feln
*gegrillt* (Gr), ge-grillt
*gehackt* (Gr), ge-hackt
*gehacktes Rindfleisch* (Gr), ge-hack-tes Rind-fleisch

*gebaktballetjes* (Du), ge-hakt-ball-et-jes

*Gehirne* (Gr), Ge-hir-ne

*geboon* (Ia), ge-hoon

*gekocht* (Gr), ge-kocht

*gekookt* (Du), ge-kookt

*gelado* (Pg), ge-la-do

*gelatin* (US), gel-a-tin

*gelato* (It), ge-la-to

*gelato alle fragole* (It), ge-la-to al-le fra-go-le

*gelato di crema* (It), ge-la-to di cre-ma

*gelbe Rüben* (Gr), gel-be R̈-ben

*geléa* (Pg-Brazil), ge-lé-a

*gelée* (Fr), ge-lée

*gelée, à la* (Fr), ge-lée, à la

*gelée de groseille* (Fr), ge-lée de gro-seille

*geléia* (Pg), ge-lé-i-a

*gêlo* (Pg), gê-lo

*gemischter Salat* (Gr), ge-misch-ter Sa-lat

*Gemüsesuppe* (Gr), Ge-mü-se-sup-pe

*genomstekt* (Sw), ge-nom-ste-kt

*genevoise, sauce* (Fr), ge-ne-voise, sauce

*génoise* (Fr), gén-oise

*Genovese, alla* (It), Ge-no-ve-se, al-la

*geoduck* (US), ge-o-duck

*geraniums* (US), ger-an-i-ums

*germ* (US), germ

*German potato salad* (US), Ger-man po-ta-to sal-ad

*gerookt* (Du), ge-rookt

*gerookte paling* (Du), ge-rookte pa-ling

*gerookte zalm* (Du), ge-rookte zalm

*geroosterd* (Du), ge-roos-terd

*geroosterd brood en jam* (Du), ge-roos-terd brood en jam

*Geröstel* (Gr), Ge-rö-stel

*Gerste* (Gr), Ger-ste

*Gerstensuppe* (Gr), Ger-sten-sup-pe

*Gervais* (Fr), Ger-vais

*geś* (Po), geś

*geschmort* (Gr), ge-schmort

*gestoofde bieten* (Du), ge-stoof-de bie-ten

*gestoofde pruimen* (Du), ge-stoof-de prui-men

*gesztenye szív* (Hu), gesz-te-nye szív

*getost* (Sw), get-ost

*Getreide* (Gr), Ge-trei-de

*gevuld* (Du), ge-vuld

*gevulde ui* (Du), ge-vul-de ui

*Gewürze* (Gr), Ge-wür-ze

*Gewürzkuchen* (Gr), Ge-würz-ku-chen

*ghai ka gosht* (Ia), ghai ka gosht

*ghala* (Gk), gha-la

*ghalopula* (Gk), gha-lo-pu-la

*gharidhes* (Gk), gha-ri-dhes

*ghee* (Ia), ghee

*gherkin* (US), gher-kin

*ghiacciata* (It), ghiac-cia-ta

*ghiaccio* (It), ghiac-cio

*ghianda* (It), ghi-an-da

*ghiveci călugăresc* (Ru), ghi-ve-ci că-lu-gă-resc

*ghleeko* (Gk), ghlee-ko

*ghurunopulo tu ghalaktos* (Gk), ghu-ru-no-pu-lo tu gha-lak-tos

*giardiniera* (It), giar-di-nier-a

*gibanica* (SC), gi-ba-ni-ca

*gibelotte* (Fr), gi-be-lotte

*gibier* (Fr), gi-bier

*giblets* (US), gib-lets

*gibnah* (Ar), gib-nah

*gigot d'agneau* (Fr), gi-got d'agn-eau

*gigue de chevreuil* (Fr), gi-gue de chev-reuil

*gingembre* (Fr), gin-gem-bre

*ginger* (US), gin-ger

*gingerbread* (US), gin-ger-bread

*gingersnaps* (US), gin-ger-snaps

*gingili* (Ia), gin-gi-li

*gingko nuts* (US), ging-ko nuts

*ginjal* (In), gin-jal

*ginnan* (Jp), gin-nan

*ginseng* (Ch), gin-seng

*gioulbassi* (Gk), gioul-bas-si

*giovetsi* (Gk), gio-ve-tsi

*girolle* (Fr), gi-rolle

*gjedde* (Nw), gjed-de

*Gjetost* (Nw), Gjet-ost

*glace* (Fr), glace

*glacé* (US), gla-cé

*glace aux noix* (Fr), glace aux noix

*glace de viande* (Fr), glace de vi-ande

*glaces tous parfums* (Fr), gla-ces tous par-fums

*gladzica* (Po), gla-dzi-ca

*glasmästarsill* (Sw), glas-mäs-tar-sill

*glasswort* (GB), glass-wort

*Glattbutt* (Gr), Glatt-butt

*Glattroche* (Gr), Glatt-ro-che

*glaze* (US), glaze

*glaze, to* (US), glaze, to

*glazirovannye gretskie orekhi* (Rs), gla-zi-ro-van-nye gret-skie o-re-khi

*gljive* (SC), glji-ve

*globe artichoke* (US), globe ar-ti-choke

*glögg* (Sw), glögg

*Glücksschweinchen* (Gr), Glücks-schwein-chen

*glucose* (US), glu-cose

*Glühwein* (Gr), Glüh-wein

*gluten* (US), glu-ten

*gluten-free* (US), glu-ten-free

*glutinous rice flour* (US), glu-ti-nous rice flour

*glycerine* (US), glyc-er-ine

*glykadakia* (Gk), gly-ka-da-kia

*gnocchi* (It), gnoc-chi

*gnocchi di patate al sugo di pomodor* (It), gnoc-chi di pa-ta-te al su-go di po-mo-dor

*gô* (Jp), gô

*gobo* (Jp), go-bo

*godiveau* (Fr), go-di-veau

*gogol'-gogol'* (Rs), go-gol'-go-gol'

*gohan* (Jp), go-han

**gobanmono** (Jp), go-han-mo-no
**goi** (Vt), go-i
**goi cuon** (Vt), go-i cu-on
**goi gia** (Vt), go-i gi-a
**gôjiru** (Jp), gô-ji-ru
**golabki** (Po), go-lab-ki
**golden needles** (US), golden nee-dles
**golden syrup** (GB), gold-en syr-up
**golub** (SC), go-lub
**golubtsy** (Rs), go-lub-tsy
**goma** (Jp), go-ma
**goma abura** (Jp), go-ma a-bu-ra
**goma-sembei** (Jp), go-ma-sem-bei
**gomba** (Hu), gom-ba
**gombaleves** (Hu), gom-ba-le-ves
**gombamártás** (Hu), gom-ba-már-tás
**gombás hús** (Hu), gom-bás hús
**gombás palacsinták** (Hu), gom-bás pa-la-csin-ták
**gombóc** (Hu), gom-bóc
**gomoku goban** (Jp), go-mo-ku go-han
**goober** (US), goo-ber
**goose** (US), goose
**gooseberry** (US), goose-ber-ry
**goosefish** (US), goose-fish
**gorditas** (Mx), gor-di-tas
**gorduroso** (Pg), gor-du-ro-so
**goreng** (In), go-reng
**goreng babat asam pedas** (In), go-reng ba-bat a-sam pe-das
**goreng cumi-cumi** (In), go-reng cu-mi-cu-mi
**Gorgonzola** (It), Gor-gon-zo-la
**goriachii shokolad** (Rs), gor-ia-chii sho-ko-lad
**goriachii vinegret iz teliachikh mozgov** (Rs), gor-ia-chii vi-ne-gret iz tel-ia-chikh moz-gov
**gorp** (US), gorp
**gosht badaam pasanda** (Ia), gosht ba-daam pa-san-da
**gosht do pyaza** (Ia), gosht do pya-za
**goshtaba** (Ia), gosht-a-ba
**gosht quorma** (Ia), gosht quor-ma
**Götterspeise** (Gr), Göt-ter-spei-se
**Gouda** (Du), Gou-da
**gougères** (Fr), gou-gères
**goujon** (Fr), gou-jon
**goulash** (US), gou-lash
**gourd** (US), gourd
**gourre** (Fr), gourre
**goûter** (Fr), goû-ter
**govedina** (SC), go-ve-di-na
**govedje pečenje** (SC), go-ve-dje pe-če-nje
**govedska juha** (SC), go-ved-ska ju-ha
**goviadina** (Rs), go-via-di-na
**goviadina po-Gusarski** (Rs), go-via-di-na po-Gu-sar-ski
**goviadina po-Stroganovski** (Rs), go-via-di-na po-Stro-ga-nov-ski

*goviazhii kotlety* (Rs), go-via-zhii kot-le-ty

*grädde* (Sw), gräd-de

*gräddost* (Sw), grädd-ost

*grah* (SC), grah

*graham cracker* (US), gra-ham crack-er

*graham flour* (US), gra-ham flour

*grahamkorputs* (Fi), gra-ham-kor-puts

*grain* (US), grain

*grain café* (Fr), grain ca-fé

*grain de poivre* (Fr), grain de poivre

*gram* (Ia), gram

*grana* (It), gra-na

*granada* (Sp), gra-na-da

*granadillas* (Sp), gra-na-di-llas

*Granatapfel* (Gr), Gra-nat-ap-fel

*granchio* (It), gran-chio

*grand-duc, au* (Fr), grand-duc, au

*grandmère* (US), grand-mère

*grandmère, à la* (Fr), grand-mère, à la

*grand venure* (Fr), grand ve-nure

*grani di carvi* (It), gra-ni di car-vi

*granita* (It), gra-ni-ta

*granita al limone* (It), gra-ni-ta al li-mo-ne

*granité* (Fr), gran-i-té

*granité au champagne* (Fr), gran-i-té au cham-pa-gne

*Granny Smith* (US), Gran-ny Smith

*grano* (It), gra-no

*granola* (US), gran-o-la

*grano saraceno* (It), gra-no sa-ra-ce-no

*granulated sugar* (US), gran-u-lat-ed sug-ar

*grãos* (Pg), grã-os

*grãos com espinafres* (Pg), grã-os com es-pi-na-fres

*grapefruit* (US), grape-fruit

*grapefrukt* (Nw), grape-frukt

*grape must bread* (US), grape must bread

*grapes* (US), grapes

*grape sugar* (US), grape sug-ar

*grappa* (It), grap-pa

*grašak* (SC), gra-šak

*gras-double* (Fr), gras-dou-ble

*gräslök* (Sw), gräs-lök

*grătar amestecat* (Ro), gră-tar a-mes-te-cat

*grate, to* (US), grate, to

*gratin, au* (Fr), gra-tin, au

*gratin dauphinois* (Fr), gra-tin dau-phi-nois

*gratin de poires* (Fr), gra-tin de poires

*gratinée* (Fr), gra-ti-née

*grauwe poon* (Du), grau-we poon

*gravlax* (Sw), grav-lax

*gravlaxsås* (Sw), grav-lax-sås

*gravmakrell* (Nw), grav-ma-krell

*gravy* (US), gra-vy

**grechnevaia kasha** (Rs), grech-ne-va-ia ka-sha

**grecque, à la** (Fr), grecque, à la

**green cauliflower** (US), green cau-li-flow-er

**greengage plum** (US), green-gage plum

**green onions** (US), green on-ions

**green pepper** (US), green pep-per

**greens** (US), greens

**green sauce** (US), green sauce

**green tea** (US), green tea

**greipin** (Fi), gre-i-pin

**greippimehu** (Fi), gre-ip-pi-me-hu

**grejpfrut** (Po), grejp-frut

**grelhado** (Pg), gre-lha-do

**grelos** (Pg), gre-los

**grenade** (Fr), gre-nade

**grenadine** (US), gren-a-dine

**grenoblois** (Fr), gre-nob-lois

**grenouille** (Fr), gre-nou-ille

**grenouilles à la provencale** (Fr), gre-nou-illes à la pro-ven-cale

**grepfrut sok** (SC), grep-frut sok

**gresskar** (Nw), gress-kar

**gribi v smetane** (Rs), gri-bi v sme-ta-ne

**gribnaia ikra** (Rs), grib-na-ia i-kra

**gribnye kotlety** (Rs), grib-nye kot-le-ty

**griddle** (US), grid-dle

**griddle cake** (US), grid-dle cake

**Griessmehl** (Gr), Griess-mehl

**griglia, alla** (It), grig-lia, al-la

**grigliata di verdure** (It), grig-lia-ta di ver-dure

**grill** (US), grill

**grillade panée** (Fr), gri-llade pa-née

**grillat** (Sw), gril-lat

**grillattu** (Fi), gril-lat-tu

**grillé** (Fr), gri-llé

**grind, to** (US), grind, to

**grinder** (US), grind-er

**gris koh** (SC), gris koh

**grissini** (It), gris-si-ni

**grits** (US), grits

**groat** (US), groat

**groenten** (Du), groen-ten

**groentensoep** (Du), groen-ten-soep

**gröna bönor** (Sw), grö-na bö-nor

**grønlangkål** (Da), grøn-lang-kål

**grønnsaker** (Nw), grønn-sak-er

**grønnsaksuppe** (Nw), grønn-sak-supp-e

**grøntsager** (Da), grønt-sag-er

**grönsakssoppa** (Sw), grön-saks-sop-pa

**grönsallad** (Sw), grön-sal-lad

**groseille à maquereau** (Fr), gro-seille à ma-que-reau

**groseille blanche** (Fr), gro-seille blanche

**groseille rouge** (Fr), gro-seille rouge

**grosse Bohne** (Gr), gros-se Boh-ne

**groszek zielony** (Po), gros-zek zie-lo-ny

**grøt** (Nw), grøt

**gröt** (Sw), gröt

**ground cherry** (US), ground cher-ry

**groundnut** (US), ground-nut

**grouper** (US), grou-per

**grouse** (US), grouse

**grovbrød** (Nw), grov-brød

**groždje** (SC), grož-dje

**gruel** (GB), gruel

**grüne Bohnen** (Gr), grü-ne Boh-nen

**grüne Paprikaschoten** (Gr), grü-ne Pap-ri-ka-scho-ten

**grüner Salat** (Gr), grü-ner Sa-lat

**Grunkern** (Gr), Grun-kern

**grunt** (US), grunt

**grunt** (US), grunt

**Gruyère** (Fr-Swiss), Gru-yère

**grystek** (Sw), gry-stek

**grzanka** (Po), grzan-ka

**grzyby** (Po), grzy-by

**guacamole** (Mx), gua-ca-mo-le

**guajalote** (Mx), gua-ja-lo-te

**guanabana** (Cb), gua-na-ba-na

**guarapo de caña** (Sp-Puerto Rico), gua-ra-po de ca-ña

**guasacaca** (Sp), gua-sa-ca-ca

**guava** (US), gua-va

**guava duff** (Cb), gua-va duff

**guay tiaw** (Th), guay ti-aw

**guazzetto** (It), guaz-zet-to

**gubana** (It), gu-ba-na

**gudeg** (In), gu-deg

**gudgeon** (GB), gudg-eon

**Gugelhupf** (Gr), Gug-el-hupf

**guinea hen** (US), guin-ea hen

**guisado** (Pg), gui-sa-do

**guisantes a la bilbaina** (Sp), gui-san-tes a la bil-ba-i-na

**guiso** (Sp), gui-so

**guiso de atun** (Sp), gui-so de a-tun

**guiso de quimbombo** (Cb), gui-so de quim-bom-bo

**gula** (In, Ml), gu-la

**gulab jamun** (Ia), gu-lab ja-mun

**gulai** (In, Ml), gu-lai

**gulai bagar** (In), gu-lai ba-gar

**gulai daging lembu** (Ml), gu-lai da-ging lem-bu

**gulai ikan padang** (In), gu-lai i-kan pa-dang

**gulai otak** (In), gu-lai o-tak

**gul ärtsoppa** (Sw), gul ärt-sop-pa

**guláš** (Cz), gu-láš

**gulášová polévka** (Cz), gu-lá-šo-vá po-lév-ka

**gule ærter** (Da), gu-le ær-ter

**guleng** (Ml), gu-leng

**gulerødder** (Da), gu-le-rød-der

**gullflyndre** (Nw), gull-flyn-dre

**gulrøtter** (Nw), gul-røt-ter

**gulýs** (Hu), gu-lyás

**gulyásleves** (Hu), gu-lyás-le-ves

**gum arabic** (US), gum ar-a-bic

**gumbo** (US), gum-bo

**gundel** (Hu), gun-del

**gunpowder tea** (US), gun-pow-der tea

**gur** (Ia), gur

**gurami** (Th), gu-ra-mi

**gurda** (Ia), gur-da

**guriev kasha** (Rs), gur-iev ka-sha

**gurka** (Sw), gur-ka

**Gurken** (Gr), Gur-ken

**Gurkensalat** (Gr), Gur-ken-sa-lat

**gurnard** (US), gur-nard

**guska** (SC), gu-ska

**güveç** (Tr), gü-veç

**gvina** (Jw-Israel), g-vi-na

**gwangen** (GB-Wales), gwan-gen

**gwaytio** (Th), gway-tio

**gwyniad y môr** (GB-Wales), gwyn-iad y môr

**gyokai ryōri** (Jp), gy-o-kai ry-ō-ri

**gyokuro** (Jp), gy-o-ku-ro

**gyro** (Gk), gy-ro

**gyümölcslé** (Hu), gyü-mölcs-lé

**gyümölcssaláta** (Hu), gyü-mölcs-sa-lá-ta

**gyūnabe** (Jp), gyū-na-be

**gyūniku** (Jp), gyū-ni-ku

**gyūniku no yawatamaki** (Jp), gyū-ni-ku no ya-wa-ta-ma-ki

# *H*

**baaievinnen sop** (In), haaie-vin-nen sop

**baarukkaleivät** (Fi), haa-ruk-ka-lei-vät

**baas** (Du), haas

**Haas avocado** (US), Haas a-vo-ca-do

**baba** (Sp), ha-ba

**babas con longanizas** (Sp), ha-bas con lon-ga-ni-zas

**babas verdes** (Sp), ha-bas ver-des

**babbah sawda** (Ar), hab-bah saw-da

**bacher** (Fr), ha-cher

**bachis** (Da), ha-chis

**backat** (Sw), ha-ckat

**backat kött** (Sw), ha-ckat kött

**Hackbraten** (Gr), Hack-bra-ten

**Hackrahmsteak** (Gr), Hack-rahm-steak

**haddock** (US), had-dock

**haggamuggie** (Sc), hag-ga-mug-gie

**haggis** (Sc), hag-gis

**hagyma** (Hu), hagy-ma

**hagymamártás** (Hu), hagy-ma-már-tás

**hái jè pí** (Ch), hái jè pí

**hăi-shēn** (Ch), hăi-shēn

**haiver sin patladjan** (Bu), haiv-er sin pat-la-djan

**hăi-wèi shālà** (Ch), hăi-wèi shā-là

**hai yup yue** (Ml), hai yup yue

**hake** (US), hake

**hakkebøf med løg** (Da), hak-ke-bøf med løg

**bakket kjøtt** (Nw), hak-ket kjøtt

**hakusai** (Jp), ha-ku-sai

**hakusai no shizuke** (Jp), ha-ku-sai no shi-zu-ke

**halal** (Ar), ha-lal

**halászlé** (Hu), ha-lász-lé

**haldi** (Ia), hal-di

**haldi jhinga** (Ia), hal-di jhin-ga

**half-and-half** (US), half-and-half

**half shell** (US), half shell

**halibut** (US), hal-i-but

**hallacas** (Sp), ha-lla-cas

**hälleflundra** (Sw), häl-le-flun-dra

**hallon** (Sw), hal-lon

**Haloumi** (Gk), Ha-lou-mi

**hälsingeost** (Sw), häl-sing-e-ost

**haluwa** (Af-Swahili), ha-lu-wa

**halva** (US), hal-va

**halva** (Gk), hal-va

**ham** (US), ham

**hamachi** (Jp), ha-ma-chi

**hamachi tataki** (Jp), ha-ma-chi ta-ta-ki

**hamaguri** (Jp), ha-ma-gu-ri

**hamaguri sakani** (Jp), ha-ma-gu-ri sa-ka-ni

**hamaguri ushiojiru** (Jp), ha-ma-gu-ri u-shio-ji-ru

**hamantasch** (Jw), ha-man-tasch

**hamburger** (US), ham-burg-er

**ham hock** (US), ham hock

**Hammelbraten** (Gr), Ham-mel-bra-ten

**Hammelfleisch** (Gr), Ham-mel-fleisch

**Hammelkeule** (Gr), Ham-mel-keu-le

**Hammelkotelett** (Gr), Ham-mel-ko-te-lett

**hamur** (Ar), ha-mur

**hanagatsuo** (Jp), ha-na-gat-su-o

**hand pies** (US), hand pies

**handsel** (Sc), hand-sel

**Hangtown Fry** (US), Hang-town Fry

**hanhi** (Fi), han-hi

**hanim göbĕgi** (Tr), ha-nim gö-bĕgi

**hanjuku tamago** (Jp), han-ju-ku ta-ma-go

**hanne tamago** (Jp), han-ne ta-ma-go

*hapanimeläkastike* (Fi), ha-pan-i-me-lä-kas-ti-ke
*hapankaalikeitto* (Fi), ha-pan-kaa-li-ke-it-to
*hapankerma* (Fi), ha-pan-ker-ma
*hapankermakastike* (Fi), ha-pan-ker-ma-kas-ti-ke
*hapanleipä* (Fi), ha-pan-lei-pä
*hapu'u* (Pl), ha-pu'u
*hara dhania* (Ia), ha-ra dha-nia
*bäränhäntäliemi* (Fi), hä-rän-hän-tä-lie-mi
*hardal* (Tr), har-dal
*hardali* (Af-Swahili), har-da-li
*hard gekookte eieren* (Du), hard ge-kook-te ei-e-ren
*hårdkogte æg* (Da), hård-kog-te æg
*hårdkokta ægg* (Sw), hård-kok-ta ægg
*Härdöpfelstock* (Gr-Swiss), Härd-öp-fel-stock
*hard sauce* (US), hard sauce
*hardshell clams* (US) hard-shell clams
*hardtack* (US), hard-tack
*hare* (US), hare
*haree mirch* (Ia), ha-ree mirch
*hareng* (Fr), har-eng
*hareng saur* (Fr), har-eng saur
*haresteg* (Da), ha-re-steg
*haricots* (Fr), ha-ri-cots
*haricots de mouton* (Fr), ha-ri-cots de mou-ton

*haricots de Soissons* (Fr), ha-ri-cots de Sois-sons
*haricots flageolets* (Fr), ha-ri-cots fla-geo-lets
*haricots larges* (Fr), ha-ri-cots lar-ges
*haricots verts* (Fr), ha-ri-cots verts
*haricots verts à la maitre d'hôtel* (Fr), ha-ri-cots verts à la maitre d'hô-tel
*haricots verts sautés au beurre* (Fr), ha-ri-cots verts sau-tés au beurre
*harina de maíz* (Sp), ha-ri-na de ma-íz
*haring* (Du), ha-ring
*haringa* (SC), ha-rin-ga
*haringsla* (Du), ha-ring-sla
*harira* (Ar), ha-ri-ra
*harissa* (Ar), ha-ris-sa
*Hartford election cake* (US), Hart-ford e-lec-tion cake
*hartgekochte Eier* (Gr), hart-ge-koch-te Ei-er
*harusame* (Jp), ha-ru-sa-me
*Harvard beets* (US), Har-vard beets
*Hase* (Gr), Ha-se
*Haselnussrahm* (Gr), Ha-sel-nuss-rahm
*Hasenkeule* (Gr), Ha-sen-keu-le
*Hasenpfeffer* (Gr), Ha-sen-pfef-fer
*Hasenrücken* (Gr), Ha-sen-rüc-ken
*hash* (US), hash
*hashed brown potatoes* (US), hashed brown po-ta-toes

*hasselnötter* (Sw), ha-ssel-nö-tter
*hasselnøtter* (Nw), hass-el-nøtt-er
*hasty pudding* (US), hast-y pud-ding
*hati ayam asam manis* (In), ha-ti a-yam a-sam ma-nis
*háu* (Ch), háu
*haupia* (Pl), ha-u-pi-a
*Häuptelsalat* (Gr-Austria), Häup-tel-sa-lat
*haute cuisine* (Fr), haute cui-sine
*háu-yóu* (Ch), háu-yóu
*háu-yóu mwó-gŭ* (Ch), háu-yóu mwó-gŭ
*Havarti* (Da), Ha-var-ti
*havregrød* (Da), hav-re-grød
*havregrøt* (Nw), hav-re-grøt
*havremel* (Nw), hav-re-mel
*havuç* (Tr), ha-vuç
*havyar* (Tr), hav-yar
*haw mok* (Th), haw mok
*hazelnut* (US), ha-zel-nut
*headcheese* (US), head-cheese
*head lettuce* (US), head let-tuce
*heart of palm* (US), heart of palm
*heavy cream* (US), hea-vy cream
*hé-bāu dàn* (Ch), hé-bāu dàn
*hebi* (Pl), he-bi
*Hecht* (Gr), Hecht
*hedelmäkakut* (Fi), he-del-mä-ka-kut
*hedelmäkukkoset* (Fi), he-del-mä-kuk-ko-set

*Hefe* (Gr), He-fe
*Hefengebäck* (Gr), He-fen-ge-bäck
*Heidelbeere* (Gr), Hei-del-bee-re
*hēi-dòu* (Ch), hēi-dòu
*heilbot* (Du), heil-bot
*Heilbutt* (Gr), Heil-butt
*helado* (Sp), he-la-do
*hellefisk* (Nw), hel-le-fisk
*helleflynder* (Da), hel-le-fly-nder
*helt* (Da), helt
*helva* (Tr), hel-va
*hepia* (In), he-pi-a
*herbata* (Po), her-ba-ta
*herbata z mlekiem* (Po), her-ba-ta z mle-kiem
*herb butter* (US), herb but-ter
*herbes* (Fr), herbes
*herbes de Provence* (Fr), herbes de Pro-vence
*herbs* (US), herbs
*herb vinegar* (US), herb vin-e-gar
*hering* (Hu), her-ing
*Heringsalat* (Gr), Her-ing-sa-lat
*Heringskönig* (Gr), Her-ing-skö-nig
*herkkusienikeitto* (Fi), herk-ku-sie-ni-ke-it-to
*herneitä* (Fi), her-ne-i-tä
*hernekeitto* (Fi), her-ne-ke-it-to
*hero* (US), he-ro
*herrgårdsost* (Sw), herr-gårds-ost
*herring* (US), her-ring
*Hervé* (Bl), Her-vé

*Herz* (Gr), Herz
*Herzoginkartoffeln* (Gr), Her-zog-in-kar-tof-feln
*hé-táu* (Ch), hé-táu
*hideg meggyleves* (Hu), hi-deg meggy-le-ves
*hidratos de carbone* (Sp), hi-dra-tos de car-bo-ne
*hiekka kakkuja* (Fi), hiek-ka kak-ku-ja
*hielo* (Sp), hi-e-lo
*hierbabuena* (Sp), hi-erb-a-bu-e-na
*hierro* (Sp), hi-er-ro
*higashi* (Jp), hi-ga-shi
*hiivaleipä* (Fi), hii-va-le-i-pä
*hiivapannukakku* (Fi), hii-va-pan-nu-kak-ku
*hijiki* (Jp), hi-ji-ki
*hikiniku* (Jp), hi-ki-ni-ku
*hillo* (Fi), hil-lo
*hilu huli* (Pl), hi-lu hu-li
*Himbeeren* (Gr), Him-bee-ren
*Himbeergeist* (Gr), Him-beer-geist
*Himbeersaft* (Gr), Him-beer-saft
*hindbær* (Da), hind-bær
*hindi* (Tr), hin-di
*hing* (Ia), hing
*hinojo* (Sp), hi-no-jo
*hirame* (Jp), hi-ra-me
*hireniku* (Jp), hi-re-ni-ku
*Hirschbraten* (Gr), Hirsch-bra-ten
*Hirschkeule* (Gr), Hirsch-keu-le
*Hirschsteak* (Gr), Hirsch-steak

*hirvenliha* (Fi), hir-ven-li-ha
*hitashimono* (Jp), hi-ta-shi-mo-no
*hiyamugi* (Jp), hi-ya-mu-gi
*hjärta* (Sw), hjär-ta
*hjerter stegte* (Da), hjer-ter steg-te
*hjortron* (Sw), hjor-tron
*hjortstek* (Sw), hjort-stek
*hladan čaj* (SC), hla-dan čaj
*hlávkový salát* (Cz), hláv-ko-vý sa-lát
*hleb* (SC), hleb
*hoagie* (US), hoa-gie
*hochepot* (Bl), hoche-pot
*Hochwild* (Gr), Hoch-wild
*hongos* (Sp), hon-gos
*ho'i'o* (Pl), ho'i'o
*hoisin* (Ch), hoi-sin
*hollandaise, sauce* (Fr), hol-lan-daise, sauce
*Hollander Kirschtorte* (Gr), Hol-lan-der Kirsch-tor-te
*hollantilainen kastike* (Fi), hol-lan-ti-la-i-nen kas-ti-ke
*Holsteinerschnitzel* (Gr), Hol-stei-ner-schnit-zel
*holub* (Cz), ho-lub
*homar* (Hu, Po), ho-mar
*homard* (Fr), ho-mard
*homard à l'américaine* (Fr), ho-mard à l'a-mé-ri-caine
*homard à la Newburg* (Fr), ho-mard à la New-burg
*homard cardinal* (Fr), ho-mard car-di-nal
*homard grillé* (Fr), ho-mard gril-lé
*homard sautè* (Fr), ho-mard sau-tè

**homard Thermidor** (Fr), ho-mard Ther-mi-dor

**hominy** (US), hom-i-ny

**hominy grits** (US), hom-i-ny grits

**høne** (Nw), hø-ne

**hongo** (Sp), hon-go

**hongroise, à la** (Fr), hon-groise, à la

**Honig** (Gr), Ho-nig

**Honigkuchen-wurfel** (Gr), Ho-nig-ku-chen-wur-fel

**honing** (Du), ho-ning

**honingkoek** (Du), ho-ning-koek

**hønsekødsuppe** (Da), høn-se-kød-sup-pe

**høns i peberrod** (Da), høns i pe-ber-rod

**hönssoppa** (Sw), höns-sop-pa

**hontaka** (Jp), hon-ta-ka

**honu** (Pl), ho-nu

**Hopfenkäse** (Gr), Hop-fen-kä-se

**hoppin' John** (US), hop-pin' John

**hops** (US), hops

**hořčice** (Cz), hoř-či-ce

**horcheta** (Sp), hor-che-ta

**hōrensō** (Jp), hō-ren-sō

**hōrensō no goma ae** (Jp), hō-ren-sō no go-ma ae

**horiatiki** (Gk), hor-ia-ti-ki

**horké kakao** (Cz), hor-ké ka-ka-o

**horn** (Nw), horn

**horned melon** (US), horned mel-on

**hornfisk** (Da), horn-fisk

**horno, al** (Sp), hor-no, al

**hors d'oeuvre** (Fr), hors d'oeu-vre

**horseradish** (US), horse-rad-ish

**horseradish sauce** (US), horse-rad-ish sauce

**hortobágyi palacsinta** (Hu), hor-to-bá-gyi pa-la-csin-ta

**hotategai** (Jp), ho-ta-te-gai

**hot cross buns** (GB), hot cross buns

**hot dog** (US), hot dog

**houbová omáčka** (Cz), ho-u-bo-vá o-máč-ka

**houby** (Cz), ho-u-by

**houby s octem** (Cz), ho-u-by s oc-tem

**hough** (Sc), hough

**houska** (Cz), ho-u-ska

**houskový knedlík** (Cz), ho-u-sko-vý kned-lík

**hout** (Ar), hout

**hovězí maso** (Cz), ho-vě-zí ma-so

**hovězí vývar s játrovými knedlíčky** (Cz), ho-vě-zí vý-var s ját-ro-vý-mi kned-líč-ky

**howtowdie** (Sc), how-tow-die

**brachová polévka** (Cz), hra-cho-vá po-lév-ka

**brášek** (Cz), hrá-šek

**brozny vina** (Cz), hroz-ny vi-na

**bruška** (Cz), hruš-ka

**buachinango** (Sp), hu-a-chi-nan-go

**Hubbard squash** (US), Hub-bard squash

*huckleberry* (US), huck-le-ber-ry
*huevas reales* (Sp), hue-vas re-a-les
*huevo* (Sp), hue-vo
*huevo duro* (Sp), hue-vo du-ro
*huevo escalfado* (Sp), hue-vo es-cal-fa-do
*huevo frito* (Sp), hue-vo fri-to
*huevo passado por agua* (Sp), hue-vo pas-sado por a-gua
*huevo revuelto* (Sp), hue-vo re-vu-el-to
*huevos rancheros* (Sp), hue-vos ran-che-ros
*huevos relleños* (Sp), hue-vos rel-le-ños
*Huhn* (Gr), Huhn
*Hühnerbrühe* (Gr), Hüh-ner-brü-he
*Hühnerfrikassee* (Gr), Hüh-ner-fri-kas-see
*Hühnerpalatschinken* (Gr-Austria), Hüh-ner-pa-lat-schin-ken
*Hühnersuppe* (Gr), Hüh-ner-sup-pe
*huile* (Fr), huile
*huile de noix* (Fr), huile de noix
*huitlachoche* (Mx), huit-la-cho-che
*huîtres* (Fr), huîtres
*hú-jyāu* (Ch), hú-jyāu
*hull* (US), hull
*hummer* (Da, Nw), humm-er
*hummer* (Sw), hum-mer
*Hummer* (Gr), Hum-mer

*hummeri* (Fi), hum-me-ri
*hummus* (Ar), hum-mus
*humr* (Cz), hum-r
*hunajaluumut* (Fi), hu-na-ja-luu-mut
*hunajapiima* (Fi), hu-na-ja-pii-ma
*hunayni* (Ar), hu-nay-ni
*hundred-year eggs* (Ch), hun-dred-year eggs
*húng chá* (Ch), húng chá
*húng-lwó-bwō* (Ch), húng-lwó-bwō
*húng-shāu li-yú* (Ch), húng-shāu li-yú
*húng-syí-lwó-bwó* (Ch), húng-syí-lwó-bwó
*húng-tsài-tóu* (Ch), húng-tsài-tóu
*hünkâr beğendi* (Tr), hün-kâr be-ğen-di
*hurka* (Hu), hur-ka
*husa* (Cz), hu-sa
*húsgombóc* (Hu), hús-gom-bóc
*hush puppy* (US), hush pup-py
*husí játra* (Cz), hu-sí ját-ra
*hutspot met klapstuk* (Du), hut-spot met klap-stuk
*huzarensla* (Du), hu-za-ren-sla
*hvetebrød* (Nw), hve-te-brød
*hvetemel* (Nw), hve-te-mel
*hvidløget* (Da), hvid-løg-et
*hvid sagosuppe* (Da), hvid sa-go-sup-pe
*hvitkål* (Nw), hvit-kål
*hvitløk* (Nw), hvit-løk
*hvitting* (Nw), hvit-ting

*hwáng-dòu* (Ch), hwáng-dòu
*hwáng-gwā* (Ch), hwáng-gwā
*hwáng-yú* (Ch), hwáng-yú
*hwā-shēng* (Ch), hwā-shēng
*hwā-shēng-yóu* (Ch), hwā-shēng-yóu
*hwe dup bup* (Kr), hwe dup bup
*hwéi-syāng-dz* (Ch), hwéi-syāng-dz
*hwo-jī* (Ch), hwo-jī

*hwún-dwùn* (Ch), hwún-dwùn
*hyacinth bean* (US), hy-a-cinth bean
*hydrates de carbone* (Fr), hy-dra-tes de car-bone
*hydrolyzed vegetable protein* (US), hy-dro-lyzed veg-e-ta-ble pro-tein
*hyldebærsuppe* (Da), hyl-de-bær-sup-pe

# I

*iablochnyi pirog* (Rs), ia-bloch-nyi pi-rog
*iabloki v kreme* (Rs), ia-blo-ki v kreme
*iadritsa* (Rs), ia-dri-tsa
*iaichnyi pashtet* (Rs), ia-ich-nyi pash-tet
*i'alawalu* (Pl-Hawaii), i'al-a-wa-lu
*iantarnaia ukha* (Rs), ian-tar-na-ia u-kha
*ice* (US), ice
*iceberg lettuce* (US), ice-berg let-tuce
*ice cream* (US), ice cream
*ices* (US), ic-es
*ichiban dashi* (Jp), i-chi-ban da-shi

*ichigo* (Jp), i-chi-go
*icicle pickles* (US), i-ci-cle pick-les
*icing* (US), ic-ing
*icing sugar* (GB), ic-ing sug-ar
*iç pilav* (Tr), iç pi-lav
*idli* (Ia), id-li
*idrati di carbonio* (It), i-dra-ti di car-bo-nio
*iers mos* (Du), i-ers mos
*igaguri* (Jp), i-ga-gu-ri
*igname* (Fr), i-gna-me
*iguana* (US), i-gua-na
*ijs* (Du), ijs
*ika* (Jp), i-ka
*ika maruyaki* (Jp), i-ka ma-ru-ya-ki

**ikan asam manis** (In), i-kan
a-sam ma-nis
**ikan asap** (In), i-kan a-sap
**ikang** (Ml), i-kang
**ikan goreng** (In), i-kan go-
reng
**ikan gurita** (In), i-kan gu-ri-
ta
**ikan masak kelapa** (In),
i-kan ma-sak ke-la-pa
**ikan pangeh** (In), i-kan pangeh
**ikra** (Rs), i-kra
**ikra iz baklazhanov** (Rs),
i-kra iz bak-la-zha-nov
**ikura** (Jp), i-ku-ra
**île flottante** (Fr), île flot-tante
**ilmalish shurb** (Ar), il-ma-
lish shurb
**Imam bayildi** (Tr), I-mam
bay-il-di
**imbiss** (Gr), im-biss
**imellettyperunasoselaatik-
ko** (Fi), i-mel-let-ty-pe-ru-
na-so-se-laa-tik-ko
**imli** (Ia), im-li
**imli chatni** (Ia), im-li chat-ni
**imo-sembei** (Jp), i-mo-sem-
bei
**impanato** (It), im-pa-na-to
**impériale, à l'** (Fr), im-pé-ri-
ale, à l'
**imposata** (It), im-po-sa-ta
**imu** (Pl), i-mu
**inab** (Ar), i-nab
**inarizushi** (Jp), i-na-ri-zu-shi
**in bianco** (It), in bian-co
**incaciatura** (It), in-ca-cia-tu-
ra

**incanestrato** (It), in-ca-ne-
stra-to
**in carrozza** (It), in ca-rroz-za
**incir** (Tr), in-cir
**incise, to** (US), in-cise, to
**incomplete protein** (US), in-
com-plete pro-tein
**inconnu** (Ca), in-con-nu
**indad** (Ia), in-dad
**indiána fánk** (Hu), in-di-á-
na fánk
**Indian cress** (US), In-di-an
cress
**Indian date** (US), In-di-an
date
**Indianerkrapfen** (Gr), In-di-
a-ner-krap-fen
**indiánky** (Cz), in-di-án-ky
**Indian pudding** (US), In-di-
an pud-ding
**Indian rice** (US), In-di-an
rice
**Indian shuck bread** (US),
In-di-an shuck bread
**Indian tea** (US), In-di-an tea
**indienne, à l'** (Fr), in-dienne,
à l'
**indiva del Belgia** (It), in-di-
va del Bel-gia
**induction** (US), in-duc-tion
**indushka** (Rs), in-dush-ka
**infuse, to** (US), in-fuse, to
**ingberlach** (Jw), ing-ber-lach
**ingefær** (Nw), ing-e-fær
**ingefær brød** (Da), in-ge-fær
brød
**ingefära** (Sw), in-ge-fä-ra
**ingemaakte rog** (Fi), in-ge-
maak-te rog

**ingen** (Jp), in-gen
**Ingwer** (Gr), Ing-wer
**inhame** (Pg), i-nha-me
**injera** (Af), in-je-ra
**inkfish** (US), ink-fish
**inky pinky** (Sc), ink-y pink-y
**inlagd fisk** (Sw), in-lagd fisk
**inlagd gurka** (Sw), in-lagd
  gur-ka
**inlagd sill** (Sw), in-lagd sill
**in padèlla** (It), in pa-dèl-la
**insalata** (It), in-sa-la-ta
**insalata condita** (It), in-sa-
  la-ta con-di-ta
**insalata di carciofi crudi**
  (It), in-sa-la-ta di car-cio-fi
  cru-di
**insalata di legumi** (It), in-sa-
  la-ta di le-gu-mi
**insalata di melanzane** (It),
  in-sa-la-ta di me-lan-za-ne
**insalata mista** (It), in-sa-la-
  ta mis-ta
**insalata nizzarada** (It), in-
  sa-la-ta niz-za-ra-da
**insalata russa** (It), in-sa-la-ta
  rus-sa
**insalata verde** (It), in-sa-la-
  ta ver-de
**instant food** (US), in-stant
  food
**intercostata di manzo** (It),
  in-ter-cos-ta-ta di man-zo
**interlard** (US), in-ter-lard
**intestines** (US), in-tes-tines
**intingolo** (It), in-ti-ngo-lo
**intinto** (It), in-tin-to
**in umido** (It), in u-mi-do
**invasto** (It), in-vas-to

**Inverness gingerbread** (Sc),
  In-ver-ness gin-ger-bread
**invert sugar** (US), in-vert
  sug-ar
**involti di cacio** (It), in-vol-ti
  di ca-cio
**involtini** (It), in-vol-ti-ni
**involtini di vitello** (It), in-
  vol-ti-ni di vi-tel-lo
**iodine** (US), i-o-dine
**iodized salt** (US), i-o-dized
  salt
**iota friulana** (It), io-ta fri-u-
  la-na
**ipikike kiasi** (Af-Swahili),
  i-pi-ki-ke ki-a-si
**ipikike sava** (Af-Swahili),
  i-pi-ki-ke sa-va
**irachi thoran** (Ia), i-ra-chi
  tho-ran
**Irish coffee** (US), I-rish cof-
  fee
**Irish moss** (US), I-rish moss
**Irish soda bread** (US), I-rish
  so-da bread
**Irish stew** (US), I-rish stew
**iritamago** (Jp), i-ri-ta-ma-go
**irmik çorbasi** (Tr), ir-mik
  çor-ba-si
**iron** (US), i-ron
**irradiation** (US), ir-ra-di-a-
  tion
**is** (Da, Nw, Sw), is
**iscas à portuguesa** (Pg), iscas à por-tu-gue-sa
**Ischlertörtchen** (Gr-Austrian),
  Isch-ler-tört-chen
**ise ebi** (Jp), i-se e-bi
**ishikarinabe** (Jp), i-shi-ka-ri-
  na-be

*ising* (Da), is-ing
*isinglass* (US), i-sin-glass
*isipikike sava* (Af-Swahili), i-si-pi-ki-ke sa-va
*iskaffe* (Sw), is-kaf-fe
*işkembe nobutlu* (Tr), iş-kem-be no-hut-lu
*Islay* (Sc), Is-lay
*isobe tamago yaki* (Jp), i-so-be ta-ma-go ya-ki
*isobe zukuri* (Jp), i-so-be zu-ku-ri
*iso kuninkaankala* (Fi), is-o ku-nin-kaan-ka-la
*ispanak* (Tr), is-pa-nak
*istákoz* (Tr), is-tá-koz
*istarke fritule* (SC), is-tar-ke fri-tu-le
*iste* (Sw), is-te

*istiridye* (Tr), is-ti-rid-ye
*isvand* (Da), is-vand
*isvatten* (Sw), is-vat-ten
*Italian dressing* (US), I-tal-ian dress-ing
*Italian meringue* (US), I-tal-ian me-ringue
*Italienne, à l'* (Fr), I-ta-lienne, à l'
*itametamono* (Jp), i-ta-me-ta-mo-no
*itik* (In), i-tik
*Ivar's Daughter* (Sc), I-var's Daugh-ter
*ivoire, à l'* (Fr), i-voire, à l'
*iwashi* (Jp), i-wa-shi
*iyi* (Tr), i-yi
*izgara* (Tr), iz-ga-ra
*iziumnye sukhariki* (Rs), iz-ium-nye su-khar-i-ki

# *J*

*jäätee* (Fi), jää-tee
*jäätellö* (Fi), jää-tel-lö
*jabali* (Sp), ja-ba-li
*jabali estofado* (Sp), ja-ba-li es-to-fa-do
*jablko* (Po), jabl-ko
*jablko* (Cz), ja-bl-ko
*jablkový závin* (Cz), ja-bl-ko-vý zá-vin
*jabugo* (Sp), ja-bu-go

*jabuka* (SC), ja-bu-ka
*jabuke u rumu* (SC), ja-bu-ke u ru-mu
*jabuticaba* (Pg), ja-bu-ti-ca-ba
*jachtschotel* (Du), jacht-scho-tel
*Jack cheese* (US), Jack cheese
*jackfruit* (Ia), jack-fruit
*Jacobsmuscheln* (Gr), Ja-cobs-mu-scheln

*já de* (Ch), já de
*jaee* (Ia), ja-ee
*jagaimo* (Jp), ja-ga-i-mo
*jägarschnitzel* (Sw), jä-gar-schnit-zel
*Jäger Art* (Gr), Jä-ger Art
*Jägerbraten* (Gr), Jä-ger-bra-ten
*Jägerhackbraten* (Gr), Jä-ger-hack-bra-ten
*jaggery* (Ia), jag-ger-y
*jagnjetina* (SC), jag-nje-ti-na
*jagnjjeća sarma u maramici* (SC), jag-nj-je-ća sar-ma u ma-ra-mi-ci
*jagntegh* (Tr), ja-gn-tegh
*jagode* (SC), ja-go-de
*jahe* (In), ja-he
*jahody* (Cz), ja-ho-dy
*jaiba* (Mx), ja-i-ba
*jaiphal* (Ia), ja-i-phal
*jailles* (Fr), ja-illes
*jaja* (SC), ja-ja
*jajegh* (Tr), ja-jegh
*jajka* (Po), jaj-ka
*jajka jajecznica* (Po), jaj-ka ja-jecz-ni-ca
*jajka na miękko* (Po), jaj-ka na miek-ko
*jajka na twardo* (Po), jaj-ka na twar-do
*jajka sadzone* (Po), jaj-ka sa-dzo-ne
*jak* (Ia), jak
*Jakobmuscheln* (Gr), Ja-kob-mu-scheln
*jalapeño* (Mx), ja-la-pe-ño
*jalea* (Sp), ja-le-a
*jalea de guayaba* (Sp), ja-le-a de gua-ya-ba

*jalebi* (Ia), ja-le-bi
*jälkiruokavohvelit* (Fi), jäl-ki-ru-o-ka-voh-ve-lit
*jallab* (Ar), jal-lab
*jalousie* (Fr), ja-lou-sie
*jam* (US), jam
*Jamaica flower* (US), Ja-mai-ca flow-er
*Jamaica pepper* (Cb), Ja-mai-ca pep-per
*jambalaya* (US), jam-ba-la-ya
*jamberry* (US), jam-ber-ry
*jambon* (Fr), jam-bon
*jambon bayonnaise* (Fr), jam-bon ba-yon-naise
*jambon cru* (Fr), jam-bon cru
*jambon de paques* (Fr), jam-bon de pa-ques
*jambon fumé* (Fr), jam-bon fu-mé
*jambon persillé* (Fr), jam-bon per-sil-lé
*jamón* (Sp), ja-món
*jamón en dulce* (Sp), ja-món en dul-ce
*jamón serrano* (Sp), ja-món ser-ran-o
*jamu* (Jp), ja-mu
*jamur* (In), ja-mur
*Jamwurzel* (Gr), Jam-wur-zel
*Jan Hagel* (Du), Jan Ha-gel
*jan in de zak* (Du), jan in de zak
*jänis* (Fi), jä-nis
*Janssonin kiusaus* (Fi), Jans-so-nin ki-u-sa-us
*Jansson's frestelse* (Sw), Jan-sson's fres-tel-se
*Jansson's temptation* (US), Jan-sson's temp-ta-tion

*jan-swèi de nyóu-ròu* (Ch), jan-swèi de nyóu-ròu

*jantung* (In), jan-tung

**Japanese cucumber** (US), Jap-a-nese cu-cum-ber

**Japanese horseradish** (US), Jap-a-nese horse-rad-ish

**Japanese pear** (US), Jap-a-nese pear

**Japanese rice** (US), Jap-a-nese rice

**Japanese white radish** (US), Jap-a-nese white rad-ish

*japonaise, à la* (Fr), ja-po-naise, à la

*jardinière, à la* (Fr), jar-di-nière, à la

*jarish* (Ar), ja-rish

*Jarlsberg* (Nw), Jarls-berg

*järpe* (Sw), jär-pe

*jarp yun yuk* (Ml), jarp yun yuk

*jarret de veau* (Fr), ja-rret de veau

*jasmine* (US), jas-mine

*jasne piwo* (Po), jas-ne pi-wo

*jäst* (Sw), jäst

*jastog* (SC), ja-stog

*játernice* (Cz), já-ter-ni-ce

*jat juk* (Kr), jat juk

*játra* (Cz), ját-ra

*játrová paštika* (Cz), já-tro-vá paš-ti-ka

*jau* (Ia), jau

*jauhelihapiiras* (Fi), ja-u-he-li-ha-pii-ras

*jauhelihapihvi* (Fi), ja-u-he-li-ha-pih-vi

*Jause* (Gr), Jau-se

*javali* (Pg), ja-va-li

*javitri* (Ia), ja-vi-tri

*jazir* (Cz), ja-zir

*jbin* (Ar), j-bin

**Jeff Davis pie** (US), Jeff Da-vis pie

*jegestea* (Hu), je-ges-te-a

*jebněčí* (Cz), je-hn-ě-čí

*jelitko* (Cz), je-lit-ko

*jelly* (US), jel-ly

*jelly roll* (US), jel-ly roll

*jemný sýr* (Cz), jem-ný sýr

*jenever* (Du), je-ne-ver

*jèn jyāng tsù* (Ch), jèn jyāng tsù

*Jerez* (Sp), Je-rez

*Jerez, al* (Sp), Je-rez, al

*jericalla* (Mx), je-ri-ca-lla

*jerky* (US), jerk-y

*jeruk* (In), je-ruk

*jeruk bali* (In), je-ruk ba-li

*jeruk kepruk* (In), je-ruk ke-pruk

*jeruk manis* (In), je-ruk ma-nis

**Jerusalem artichoke** (US), Je-ru-sa-lem ar-ti-choke

*jesiotr* (Po), je-sio-tr

*Jessica* (Fr), Jes-si-ca

*jetra* (SC), je-tra

*jewfish* (US), jew-fish

**Jewish cuisine** (Jw), Jew-ish cui-sine

*jhinga* (Ia), jhin-ga

*jhinga do-piaza* (Ia), jhin-ga do-pia-za

*jhinga ka khaja* (Ia), jhin-ga ka kha-ja

*jhinga kari* (Ia), jhin-ga ka-ri

*jhinga saag* (Ia), jhin-ga saag

*jhinga shorsha* (Ia), jhin-ga
shor-sha

* jī* (Ch), jī

*jibini* (Af-Swahili), ji-bi-ni

*jibn* (Ar), jib-n

*jibu ni* (Jp), ji-bu ni

*jibu ni yoshizen* (Jp), ji-bu ni
yo-shi-zen

*jícama* (Mx), jí-ca-ma

*jī-dàn* (Ch), jī-dàn

*jídla* (Cz), jí-dla

*jī ěr yān wō tāng* (Ch), jī ěr
yān wō tāng

*jīn-jú* (Ch), jīn-jú

*jī tāng* (Ch), jī tāng

*jiternice* (Cz), ji-ter-ni-ce

*jì-yú* (Ch), jì-yú

*jocoque* (Sp), jo-co-que

*Jod* (Gr), Jod

*Joghurt* (Gr), Jo-ghurt

*Johannisbeerkonfitüre*
(Gr), Jo-han-nis-beer-kon-
fi-tü-re

*Johannisbeerkuchen* (Gr),
Jo-han-nis-beer-ku-chen

*Johannisbrot* (Gr), Jo-han-
nis-brot

*John Dory* (US), John Do-ry

*johnny cake* (US), john-ny
cake

*joint* (US), joint

*Joinville* (Fr), Join-ville

*Joinville, sauce* (Fr), Join-
ville, sauce

*Jonchée* (Fr), Jon-chée

*jong* (Du), jong

*jong belegan* (Du), jong be-
le-gan

*Jordon almond* (US), Jordon al-mond

*jordbær* (Da, Nw), jord-bær

*jordgubbar* (Sw), jord-gub-
bar

*jordnøtter* (Nw), jord-nøtt-er

*jordnötter* (Sw), jord-nöt-ter

*joulukinkku* (Fi), jo-u-lu-
kink-ku

*joululuumukakku* (Fi), jo-u-
lu-luu-mu-kak-ku

*joulutähti* (Fi), jo-u-lu-täh-ti

*jowar* (Ia), jo-war

*jow bo yay* (Ml), jow ho yay

*jowl* (US), jowl

*joyanabe* (Jp), jo-ya-na-be

*jr-má* (Ch), jr-má

*jr-má bing* (Ch), jr-má bing

*Juan canary melon* (US),
Juan ca-nar-y mel-on

*jŭ de mài-pyàn* (Ch), jŭ de
mài-pyàn

*judías* (Sp), ju-dí-as

*judías blancas* (Sp), ju-dí-as
blan-cas

*judías escarlatas* (Sp), ju-dí-
as es-car-la-tas

*judías verdes* (Sp), ju-dí-as
ver-des

*judic* (Fr), ju-dic

*jugged* (US), jugged

*jugged hare* (US), jugged
hare

*jugo* (Sp), ju-go

*jugo de manzana* (Sp), ju-go
de man-za-na

*jugurtti* (Fi), ju-gurt-ti

*juhn kol* (Kr), juhn kol

*juive, à la* (Fr), juive, à la

*jujube* (US), ju-jube

*julekage* (Nw), ju-le-ka-ge

*jule risengrød* (Da), ju-le ris-
en-grød
*julienne* (Fr), ju-li-enne
*jultallrik* (Sw), jul-tall-rik
*jumble* (US), jum-ble
*jungjang* (Kr), jung-jang
*jungkik* (Kr), jung-kik
*juniper berries* (US), ju-ni-
per ber-ries
*junipero* (Sp), ju-ní-pe-ro
*junket* (GB), jun-ket
*junk food* (US), junk food
*jū-ròu* (Ch), jū-ròu
*jū-ròu chǎu-myàn* (Ch), jū-
ròu chǎu-myàn
*jus* (Du, Fr), jus
*jus de pomme* (Fr), jus de
pomme
*jus de viande* (Fr), jus de vi-
ande
*jussière, à la* (Fr), jus-sière, à la

*jūsu* (Jp), jū-su
*jú-swun* (Ch), jú-swun
*juusto* (Fi), juus-to
*juustokukkoset* (Fi), juus-to-
kuk-ko-set
*juusto-muna voileipä* (Fi),
juus-to-mu-na vo-i-le-i-pä
*juustoruohosipulitahna*
(Fi), juus-to-ru-o-ho-si-pu-
li-tah-na
*juustotangot* (Fi), juus-to-
tan-got
*jyān dàn* (Ch), jyān dàn
*jyāng* (Ch), jyāng
*jyāng-yóu* (Ch), jyāng-yóu
*jyǎu-dz* (Ch), jyǎu-dz
*jyē-mwò-jyàng* (Ch), jyē-
mwò-jyàng
*jyou-jī* (Ch), jyou-jī
*jyú-dz* (Ch), jyú-dz
*jyú-dz jēr* (Ch), jyú-dz jēr

# K

*kaakao* (Fi), kaa-ka-o
*kaali* (Fi), kaa-li
*kaalikeitto* (Fi), kaa-li-ke-it-
to
*kaas* (Du), kaas
*kabab* (Ia), ka-bab
*kabak* (Tr), ka-bak
*kabak dolmasi* (Tr), ka-bak
dol-ma-si

*kabak tatlisi* (Tr), ka-bak tat-
li-si
*kaban* (Rs), ka-ban
*kabayaki* (Jp), ka-ba-ya-ki
*Kabeljau* (Gr), Ka-bel-jau
*kabeljauw* (Du), ka-bel-jauw
*kabeljauwstaart* (Du), ka-
bel-jauws-taart
*kabeljo* (Sw), ka-bel-jo

**kabob** (US), ka-bob
**kabocha** (Jp), ka-bo-cha
**kabu** (Jp), ka-bu
**kabuto-age** (Jp), ka-bu-to-age
**kacang** (In), ka-cang
**kacang hijau** (In), ka-cang hi-jau
**kacang panjang** (In), ka-cang pan-jang
**kachna** (Cz), kach-na
**kachumbari** (Af-Swahili), ka-chu-mba-ri
**kacsa** (Hu), ka-csa
**kaczka** (Po), kacz-ka
**kaddu** (Ia), kad-du
**kadin budu** (Tr), ka-din bu-du
**kadin göbeği** (Tr), ka-din gö-beği
**kaeng masaman** (Th), ka-eng ma-sa-man
**kærnemælk** (Da), kær-ne-mælk
**kawrnemælkskoldskål** (Da), kær-ne-mælks-kold-skål
**kærnemælkssuppe** (Da), kær-ne-mælks-sup-pe
**kafa** (SC), ka-fa
**kāfēi** (Ch), kā-fēi
**kafes** (Gk), ka-fes
**kaffe** (Da, Nw, Sw), kaf-fe
**Kaffee** (Gr), Kaf-fee
**Kaffeekuchen** (Gr), Kaf-fee-ku-chen
**Kaffee mit Schlagobers** (Gr), Kaf-fee mit Schlag-o-bers
**kage** (Da), ka-ge
**kâğit helvasi** (Tr), kâğ-it hel-va-si
**kâğitta** (Tr), kâğ-it-ta

**kahawa** (Af-Swahili), ka-ha-wa
**kahve** (Tr), kah-ve
**kahvi** (Fi), kah-vi
**kahwa** (Ar), kah-wa
**kaibashira** (Jp), ka-i-ba-shi-ra
**kail brose** (Sc), kail brose
**kaimati** (Af-Swahili), ka-i-ma-ti
**Kaiserfleisch** (Gr), Kai-ser-fleisch
**Kaiserschmarren** (Gr-Austria), Kai-ser-schmar-ren
**kajgana** (SC), ka-j-ga-na
**kajiki** (Jp), ka-ji-ki
**kajmak** (SC), ka-j-mak
**kajsija** (SC), ka-j-si-ja
**kajzerice** (SC), ka-j-ze-ri-ce
**kakao** (Nw), ka-ka-o
**kakao** (Tr), ka-kao
**kakap** (In), ka-kap
**kakavia** (Gk), ka-kav-ia
**kake** (Nw), kak-e
**kakesoba** (Jp), ka-ke-so-ba
**kaki** (Jp), ka-ki
**kaki** (Jp), ka-ki
**kakimochi** (Jp), ka-ki-mo-chi
**kakku** (Fi), kak-ku
**kakoretsi** (Gk), ka-ko-ret-si
**kål** (Da, Nw, Sw), kål
**kalafior** (Po), ka-la-fior
**kalakeitto** (Fi), ka-la-ke-it-to
**kalakukko** (Fi), ka-la-kuk-ko
**kalamarakia tighanita** (Gk), ka-la-ma-ra-kia ti-gha-ni-ta
**kalamata olives** (US), ka-la-ma-ta ol-ives

*kalasalaatti* (Fi), ka-la-sa-laat-ti
*kalbasà saséski* (Rs), kal-ba-sà sa-sés-ki
*Kalbfleisch* (Gr), Kalb-fleisch
*Kalbsbraten* (Gr), Kalbs-bra-ten
*Kalbsfrikassee* (Gr), Kalbs-fri-kas-see
*Kalbsgulasch* (Gr), Kalbs-gu-lasch
*Kalbshaxe* (Gr), Kalbs-ha-xe
*Kalbsleber* (Gr), Kalbs-le-ber
*Kalbsrolle* (Gr), Kalbs-rol-le
*Kalbsschnitzel* (Gr), Kalbs-schnit-zel
*Kalbssteak* (Gr), Kalbs-steak
*Kaldaunen* (Gr), Kal-dau-nen
*kåldolmar* (Sw), kål-dol-mar
*kåldolmer* (Da), kål-dol-mer
*kale* (US), kale
*kaléji* (Ia), ka-lé-ji
*kalfkott* (Sw), kalf-kott
*kalfsoester* (Du), kalfs-oes-ter
*kalfsvlees* (Du), kalfs-vlees
*kali* (Ia), ka-li
*kali mirch* (Ia), ka-li mirch
*kali urad* (Ia), ka-li u-rad
*kalkas* (Rs), kal-kas
*kalkan* (Tr), kal-kan
*kalkkuna* (Fi), kalk-ku-na
*kalkoen* (Du), kal-ko-en
*kalkon* (Sw), kal-kon
*kalkun* (Nw), kal-kun
*kalops* (Sw), ka-lops
*kal rulader* (Nw), kal ru-la-der
*kalsoppa* (Sw), kal-sop-pa

*kalte pikante Sosse* (Gr), kalte pi-kan-te Sos-se
*kalte Speisen* (Gr), kalte Spei-sen
*Kaltschale* (Gr), Kalt-scha-le
*kalua pua* (Pl), ka-lu-a pu-a
*kalv* (Nw), kalv
*kalvebräss* (Sw), kalve-bräss
*kalvefilé* (Nw), kalv-e-fi-lé
*kalvehjerte* (Da), kal-veh-jer-te
*kalvekarbonade* (Da), kal-ve-kar-bo-na-de
*kalvekjøtt* (Nw), kalv-e-kjøtt
*kalvekød* (Da), kal-ve-kød
*kalvestek* (Nw), kalv-e-stek
*kalvkotlett* (Sw), kalv-kot-lett
*kamaboka* (Jp), ka-ma-bo-ka
*kamano lomi* (Pl), ka-ma-no lo-mi
*kamasu* (Jp), ka-ma-su
*kamba* (Af-Swahili), ka-mba
*kambing* (In), kam-bing
*kaminarijiru* (Jp), ka-mi-na-ri-ji-ru
*kammooniyya* (Ar), kam-moo-niy-ya
*kamo* (Jp), ka-mo
*kamo namban* (Jp), ka-mo nam-ban
*kampela* (Fi), kam-pe-la
*kamrakh* (Ia), kam-rakh
*kana* (Fi), ka-na
*kanakeitto* (Fi), ka-na-ke-it-to
*kanapki z jajkami* (Po), ka-nap-ki z jaj-ka-mi
*kanasalaatti* (Fi), ka-na-sa-laat-ti
*kanelbullar* (Sw), ka-nel-bul-lar

kāng (Ch), kāng
kangaja (Af-Swahili), ka-nga-ja
kangaroo tail soup (Aa), kan-ga-roo tail soup
kani (Jp), ka-ni
kanidōfu iridashi (Jp), ka-ni-dō-fu i-ri-da-shi
kanin (Nw, Sw), ka-nin
Kaninchen (Gr), Ka-nin-chen
kanisu (Jp), ka-ni-su
kanpyo (Jp), kan-pyo
kanten (Jp), kan-ten
kapamas (Gk), ka-pa-mas
Kapernsosse (Gr), Ka-pern-sos-se
kapormártás (Hu), ka-por-már-tás
káposzta (Hu), ká-posz-ta
káposztasaláta (Hu), ká-posz-ta-sa-lá-ta
káposztás kockák (Hu), ká-posz-tás koc-kák
káposztás rétes (Hu), ká-posz-tás ré-tes
kappa maki (Jp), kap-pa ma-ki
kapr (Cz), ka-pr
kapusniak (Po), ka-pus-niak
kapusta (Po), ka-pus-ta
kapustová polévka (Cz), ka-pus-to-vá po-lév-ka
Kapuzinerkresse (Gr), Ka-pu-zi-ner-kres-se
kara age (Jp), ka-ra age
karaciğer (Tr), ka-ra-ciğ-er
karah (Ar), ka-rah
karaj (Hu), ka-raj
karalábé (Hu), ka-ra-lá-bé

Karamel (Gr), Ka-ra-mel
karamelbudding (Da), ka-ra-mel-bud-ding
karashi (Jp), ka-ra-shi
karashina (Jp), ka-ra-shi-na
kardhal (Hu), kard-hal
Kardone (Gr), Kar-do-ne
karei (Jp), ka-re-i
kare ikan (In), ka-re i-kan
karei no sashimi (Jp), ka-re-i no sa-shi-mi
karei rikyū funamori (Jp), ka-re-i ri-kyū fu-na-mo-ri
karéla (Ia), ka-ré-la
karfiol (Hu, SC), kar-fi-ol
karibayaki (Jp), ka-ri-ba-ya-ki
karides (Tr), ka-ri-des
karidopita (Gk), ka-ri-do-pi-ta
kari-kari (Ph), ka-ri-ka-ri
karişik (Tr), ka-ri-şik
karişik etler izgara (Tr), ka-ri-şik et-ler iz-ga-ra
karjalanpaisti (Fi), kar-ja-lan-pa-is-ti
karjalanpiirakoita (Fi), kar-ja-lan-pii-ra-ko-i-ta
karnabahar (Tr), kar-na-ba-har
karnabeet makly (Ar), kar-na-beet mak-ly
karnemelk (Du), kar-ne-melk
karota (Gk), ka-ro-ta
karoti (Af-Swahili), ka-ro-ti
Karotten (Gr-Austria), Ka-rot-ten
karp (Po, Rs), karp
karpe (Nw), kar-pe

**Karpfen** (Gr), Karp-fen

**karpuz** (Tr), kar-puz

**karpyon** (Jw), karp-yon

**kartoffel** (Da), kar-tof-fel

**Kartoffel** (Gr), Kar-tof-fel

**Kartoffelgemüse** (Gr), Kar-tof-fel-ge-mü-se

**Kartoffelknödel** (Gr), Kar-tof-fel-knö-del

**kartoffelmos** (Da), kar-tof-fel-mos

**Kartoffelpuffer** (Gr), Kar-tof-fel-puf-fer

**Kartoffelpüree** (Gr), Kar-tof-fel-pü-ree

**Kartoffelsalat** (Gr), Kar-tof-fel-sa-lat

**Kartoffelsuppe** (Gr), Kar-tof-fel-sup-pe

**kaşar** (Tr), ka-şar

**Käse** (Gr), Kä-se

**Käsebrot** (Gr), Kä-se-brot

**Kasein** (Gr), Ka-sein

**Käseplatte** (Gr), Kä-se-plat-te

**kaše z pohanky** (Cz), ka-še z po-han-ky

**kasha** (Rs), ka-sha

**kasha varnishkas** (Jw), ka-sha var-nish-kas

**kashtanovi pudding** (Rs), kash-ta-no-vi pud-ding

**kasséri** (Gk), kas-sé-ri

**kasséri tiganitó** (Gk), kas-sé-ri ti-ga-ni-tó

**Kassie** (Gr), Kas-si-e

**Kassler Rippchen** (Gr), Kas-sler Ripp-chen

**kastad** (Af-Swahili), kas-tad

**kastanje** (In), kas-tan-je

**kastanjer** (Nw), ka-stan-jer

**kastike** (Fi), kas-ti-ke

**kasutera** (Jp), ka-su-te-ra

**kasviskeitto** (Fi), kas-vis-ke-it-to

**kasza** (Po), kas-za

**katai roru pan** (Jp), ka-ta-i ro-ru pan

**katjang kapri** (In), kat-jang kap-ri

**katjang kedele** (In), kat-jang ke-de-le

**katjang tanah** (In), kat-jang ta-nah

**katkarapusalaatti** (Fi), kat-ka-ra-pu-sa-laat-ti

**katkaravut** (Fi), kat-ka-ra-vut

**katrinplommon** (Sw), ka-trin-plom-mon

**katsudon** (Jp), ka-tsu-don

**katsuo** (Jp), ka-tsu-o

**katsuobushi** (Jp), ka-tsu-o-bu-shi

**kău de** (Ch), kău de

**kău myàn-bāu** (Ch), kău myàn-bāu

**kău nyóu-ròu** (Ch), kău nyóu-ròu

**kău pái-gŭ** (Ch), kău pái-gŭ

**kaurapuuro** (Fi), ka-u-ra-puu-ro

**káva** (Cz), ká-va

**kavayd** (Jw), ka-vayd

**kávé** (Hu), ká-vé

**Kaviar** (Gr), Ka-vi-ar

**kavoorya** (Gk), ka-voor-ya

**kavring** (Nw), kav-ring

**kavun** (Tr), ka-vun

**kawa** (Po), ka-wa

**kaws** (Po), kaws

**kayisi** (Tr), kay-i-si

**kaymak** (Tr), kay-mak
**kaymakli elma kompostosu** (Tr), kay-mak-li el-ma kom-pos-to-su
**kazu no ko** (Jp), ka-zu no ko
**kebab** (Tr), ke-bab
**kebeji** (Af-Swahili), ke-be-ji
**kecap** (In), ke-cap
**kĕchap** (Ml), kē-chap
**kechap** (Af-Swahili), ke-chap
**kechappu** (Jp), ke-chap-pu
**kečup** (Cz), ke-čup
**kedgeree** (GB), ked-ger-ee
**kedju** (Ml), ke-dju
**kedlubny** (Cz), ked-lub-ny
**kefalos** (Gk), ke-fa-los
**kefalotiri** (Gk), ke-fa-lo-ti-ri
**kefiiri** (Fi), ke-fii-ri
**keftedakia** (Gk), kef-te-dak-ia
**keftedes** (Gk), kef-te-des
**kebrasya** (Gk), keh-ras-ya
**keitetty** (Fi), ke-i-tet-ty
**keitto** (Fi), ke-it-to
**keju** (In), ke-ju
**kĕki** (Jp), kē-ki
**keki** (Af-Swahili), ke-ki
**keki ya matunda** (Af-Swahili), ke-ki ya ma-tu-nda
**kebap** (Tr), ke-bap
**kefal** (Tr), ke-fal
**kefir** (US), ke-fir
**keftedhakia** (Gk), kef-te-dhak-ia
**keitetty muna** (Fi), ke-i-tet-ty mu-na
**kékra** (Ia), kék-ra
**Kekse** (Gr), Kek-se
**kelapa** (In), ke-la-pa

**kelapa sayur** (Ml), ke-la-pa sa-yur
**kelp** (US), kelp
**kemény tojás** (Hu), ke-mény to-jás
**kentang goreng** (In), ken-tang go-reng
**kentang panggang** (In), ken-tang pang-gang
**kentang rebus** (In), ken-tang re-bus
**kentang tumbuk** (In), ken-tang tum-buk
**kentjoer** (Du), ken-tjo-er
**Kentucky burgoo** (US), Ken-tuck-y bur-goo
**kenyér** (Hu), ke-nyér
**kepah** (In), ke-pah
**kepiting pedas** (In), ke-pi-ting pe-das
**kerie** (In), ke-rie
**kerma** (Fi), ker-ma
**kermaviilisilakka** (Fi), ker-ma-vii-li-si-lak-ka
**Kernhem** (Du), Kern-hem
**kernmilk** (Sc), kern-milk
**kerrie kool sla** (Du), ker-rie kool sla
**kersen** (Du), ker-sen
**kerupuk** (Ml), ke-ru-puk
**kesäkeitto** (Fi), ke-sä-ke-it-to
**kesäkurpitsa** (Fi), ke-sä-kur-pit-sa
**keşkec** (Tr), keş-kec
**keşkul** (Tr), keş-kul
**ketchup** (US), ket-chup
**ketimun** (In), ke-ti-mun
**ketjap** (In), ket-jap
**ketovaia** (Rs), ke-to-va-ia
**ketsup** (Du), ket-sup

**ketumbar** (In), ke-tum-bar
**ketupat** (Ml), ke-tu-pat
**kex** (Sw), kex
**Key lime pie** (US), Key lime
  pie
**khai cheow** (Th), khai cheow
**khana** (Ia), kha-na
**kharcho** (Rs), khar-cho
**khas-khas** (Ia), khas-khas
**kheer** (Ia), kheer
**khichari** (Ia), khi-cha-ri
**khleb** (Rs), khleb
**khow muck yue** (Ml), khow
  muck yue
**khoya** (Ia), kho-ya
**khtapodhi me saltsa** (Gk),
  khta-po-dhi me salt-sa
**khubz** (Ar), khubz
**khudar** (Ar), khu-dar
**kibbeh** (Ar), kib-beh
**kibbeh bissaniyyeh** (Ar), kib-
  beh bis-sa-niy-yeh
**kibbeh nayya** (Ar), kib-beh
  nay-ya
**kibda** (Ar), kib-da
**kichel** (Jw), kich-el
**Kichererbse** (Gr), Ki-cher-
  erb-se
**kid** (US), kid
**kidney** (US), kid-ney
**kidney bean** (US), kid-ney
  bean
**kielbasa** (Po), kiel-ba-sa
**kielikampala** (Fi), ki-el-i-
  kam-pa-la
**kiflik** (Hu), kif-lik
**kiichigo** (Jp), ki-ich-i-go
**kiisseli** (Fi), kiis-se-li
**kiji** (Jp), ki-ji
**kiks** (Da), kiks

**kikujisha** (Jp), ki-ku-ji-sha
**kikutane** (Jp), ki-ku-ta-ne
**kiliç baliği** (Tr), ki-liç ba-liği
**kiliç şiş** (Tr), ki-liç şiş
**kilohaili** (Fi), ki-lo-ha-i-li
**kimchee** (Kr), kim-chee
**kimo** (Jp), ki-mo
**king, à la** (US), king, à la
**king crab** (US), king crab
**king salmon** (US), king sal-
  mon
**kinkan** (Jp), kin-kan
**kinkku** (Fi), kink-ku
**kinywaji** (Af-Swahili), ki-ny-
  wa-ji
**kip** (Du), kip
**kip aan't spit** (Du), kip aan't
  spit
**kipfel** (Hu), kip-fel
**Kipfel** (Gr), Kip-fel
**kippensoep** (Du), kip-pen-
  soep
**kipper** (GB), kip-per
**kipper, to** (Sc), kip-per, to
**kirántott hal** (Hu), ki-rán-
  tott hal
**kirántott sajt** (Hu), ki-rán-
  tott sajt
**kiraz** (Tr), ki-raz
**kirjolohi** (Fi), kir-jo-lo-hi
**Kirsch** (Gr), Kirsch
**Kirschen** (Gr), Kir-schen
**Kirschwasser** (Gr), Kirsch-
  was-ser
**kirsebær** (Da, Nw), kir-se-
  bær
**kisel** (Rs), ki-sel
**kisela voda** (SC), ki-se-la vo-
  da

kishk (Ar), kishk
kishka (Jw), kish-ka
kiss (US), kiss
kisu (Jp), ki-su
kitsune udon (Jp), ki-tsu-ne
u-don
kiwanda (Af-Swahili), ki-wa-
nda
kiwano (US), ki-wa-no
kiwifruit (US), ki-wi-fruit
kiyma (Tr), kiy-ma
kizandamono (Jp), ki-zan-
da-mo-no
kizarmiş ekmek (Tr), ki-zar-
miş ek-mek
kizarmis patlican (Tr), ki-
zar-mis pat-li-can
kizartma (Tr), ki-zart-ma
kizartmasi (Tr), ki-zart-ma-si
kjeks (Nw), kjeks
kjøtt (Nw), kjøtt
kjøttboller (Nw), kjøtt-boll-er
kjøttkaker (Nw), kjøtt-kak-er
kjøttpålegg (Nw), kjøtt-på-
legg
kjøttretter (Nw), kjøtt-ret-ter
klarbär (Sw), klar-bär
klare Kraftbrühe (Gr), klare
Kraft-brü-he
kletskoppen (Du), klet-skop-
pen
klipfisk (Da), klip-fisk
klobása (Cz), klo-bá-sa
klookva (Rs), klook-va
Klösse (Gr), Klös-se
Klosterkäse (Gr), Klos-ter-
kä-se
kluski ślaskie (Po), klus-ki
ślas-kie

knäckebröd (Sw), knä-cke-
bröd
Knackwurst (Gr), Knack-
wurst
knaidel (Jw), knaid-el
knead (US), knead
knedle (SC), kned-le
knedlik (Cz), kned-lik
knekkebrød (Nw), knek-ke-
brød
Knieslück (Gr), Knies-lück
knish (Jw), knish
Knoblauch (Gr), Knob-lauch
knockwurst (US), knock-
wurst
Knödel (Gr-Austria), Knö-
del
knoflook (Du), knof-look
Knollensellerie (Gr), Knol-
len-sel-le-rie
knuckle (US), knuck-le
kobasice (SC), ko-ba-si-ce
Kobe beef (Jp), Ko-be beef
koblihy (Cz), kob-li-hy
kōcha (Jp), kō-cha
kocsonyázott hal (Hu), ko-
cso-nyá-zott hal
kocsonyázott sertés (Hu),
ko-cso-nyá-zott ser-tés
kød (Da), kød
kodok (In), ko-dok
koek (Du), koek
koekjes (Du), koek-jes
koeksisters (Af), koek-sis-ters
koffie (Du), kof-fie
köfte (Tr), köf-te
kofté (Ia), kof-té
köfte (In), köf-te
kogt oksebryst (Da), kogt ok-
se-bryst

**kogt skinke med Madeira** (Da), kogt skin-ke med Ma-dei-ra

**kōhaku namasu** (Jp), kō-ha-ku na-ma-su

**kōhii** (Jp), kō-hi-i

**kohitsuji no niku** (Jp), ko-hi-tsu-ji no ni-ku

**Kohl** (Gr), Kohl

**Kohlenhydrat** (Gr), Koh-len-hy-drat

**kohlrabi** (US), kohl-ra-bi

**Kohlrabi** (Gr), Kohl-ra-bi

**Kohlsprossen** (Gr), Kohl-spros-sen

**koi no arai** (Jp), ko-i no a-ra-i

**koko iliyo moto** (Af-Swahili), ko-ko i-li-yo mo-to

**kokoretsi** (Gk), ko-ko-ret-si

**kokt skinke** (Nw), kokt skink-e

**kokt torsk** (Sw), kokt torsk

**kol** (In), kol

**kolač** (SC), ko-lač

**koláč** (Cz), ko-láč

**koláčky** (Cz), ko-láč-ky

**kolak labu** (In), ko-lak la-bu

**kolbász** (Hu), kol-bász

**koldebord** (Da), kol-de-bord

**koldtbord** (Nw), koldt-bord

**koliflawa** (Af-Swahili), ko-li-fla-wa

**kolja** (Sw), kol-ja

**kolje** (Nw), kol-je

**kólliva** (Gk), kól-li-va

**kolozsvári rakott káposzta** (Hu), ko-lozs-vá-ri ra-kott ká-posz-ta

**kombosta** (Gk), kom-bos-ta

**kombu** (Jp), kom-bu

**kome** (Jp), ko-me

**köménymagos leves** (Hu), kö-mény-ma-gos le-ves

**komijnekaas** (Du), ko-mi-jne-kaas

**komkommer** (Du), kom-kom-mer

**komposto** (Tr), kom-pos-to

**Kompott** (Gr), Kom-pott

**kompot z jablek** (Po), kom-pot z jab-lek

**konditorkager** (Da), kon-di-tor-ka-ger

**kongesuppe** (Nw), kon-ge-sup-pe

**Königinsuppe** (Gr), Kö-ni-gin-sup-pe

**Königsberger Klops** (Gr), Kö-nigs-ber-ger Klops

**konijn** (Du), ko-nijn

**konjac** (Sw), kon-jac

**konnyaku** (Jp), kon-nya-ku

**konyun** (Tr), kon-yun

**kool** (Du), kool

**Kopfsalat** (Gr), Kopf-sa-lat

**kopi** (In), ko-pi

**kořeněné** (Cz), ko-ře-ně-né

**koření** (Cz), ko-ře-ní

**korf** (Sw), korf

**korhelyleves** (Hu), kor-hely-le-ves

**korma** (Ia), kor-ma

**korn** (Nw), korn

**koromotsuki** (Jp), ko-ro-mo-tsu-ki

**koroptev** (Cz), ko-rop-tev

**körözött júhtúró** (Hu), kö-rö-zött júh-tú-ró

**körsbär** (Sw), körs-bär

korstjes (Du), kors-tjes
körte (Hu), kör-te
korv (Sw), korv
kørvelsuppe (Da), kør-vel-
sup-pe
kosher (Jw), ko-sher
kosher pickle (Jw), ko-sher
pick-le
kosher salt (Jw), ko-sher salt
koshihikari (Jp), ko-shi-hi-
ka-ri
koshō (Jp), ko-shō
kota (Gk), ko-ta
kota kapama (Gk), ko-ta ka-
pa-ma
kotasupa (Gk), ko-ta-su-pa
Kotelett (Gr), Ko-te-lett
kotelett (Nw), ko-te-lett
kotleti (Rs), kot-le-ti
kotleti iz rybi (Rs), kot-le-ti
iz ry-bi
kotlet pane (Tr), kot-let pa-
ne
kotlety mielone (Po), kot-le-
ty mie-lo-ne
kotlety wolowe (Po), kot-le-ty
wo-lo-we
kotopulo (Gk), ko-to-pu-lo
köttbullar (Sw), kött-bul-lar
köttfärsgrotta (Sw), kött-
färs-grot-ta
köttfärsröra (Sw), kött-färs-
rö-ra
koude schotel (Du), kou-de
scho-tel
kouféta (Gk), kou-fé-ta
koulitch (Rs), kou-litch
kourambiedes (Gk), kou-
ram-bie-des

kousa mahshi (Ar), kou-sa
mah-shi
kovaksi keitetty muna (Fi),
ko-vak-si ke-i-tet-ty mu-na
krab (Cz), krab
krabba (Sw), krab-ba
krabbe (Nw), krabb-e
Krabben (Gr), Krab-ben
kraemmerhuse med fløde-
skum (Da), kraem-mer-hu-
se med flø-de-skum
Kraftbrühe (Gr), Kraft-brü-
he
Kraftbrühe mit Ei (Gr),
Kraft-brü-he mit Ei
kräftor (Sw), kräf-tor
králík (Cz), krá-lík
krasata (Gk), kra-sa-ta
kråsesuppe (Da), krå-se-sup-
pe
krastavci (SC), kra-stav-ci
Kraut (Gr), Kraut
kreatopitta (Gk), kre-a-to-
pit-ta
Krebs (Gr), Krebs
Krebsschwänze (Gr), Krebs-
schwän-ze
kreeft (Du), kreeft
krem (Nw), krem
krém (Cz), krém
kremali (Tr), kre-ma-li
krem od vanile (SC), krem
od va-ni-le
krem şanti (Tr), krem şan-ti
Kren (Gr-Austria), Kren
krentenbrood (Du), kren-
ten-brood
krentenbroodjes (Du), kren-
ten-brood-jes
kreplach (Jw), krep-lach

*kreps* (Nw), kreps
*krimu* (Af-Swahili), kri-mu
*kringle* (Nw), kring-le
*kringlor* (Sw), kring-lor
*kritharaki* (Gk), kri-tha-ra-ki
*krmenadle* (SC), kr-me-na-dle
*krocan* (Cz), kro-can
*Krokette* (Gr), Kro-kette
*kromeski* (Po), kro-mes-ki
*kromkake* (Nw), krom-ka-ke
*krompir* (SC), krom-pir
*kronärtskocka* (Sw), kron-ärts-ko-cka
*kroppkakor* (Sw), kropp-ka-kor
*kruh* (SC), kruh
*kruidkoek* (Du), kruid-koek
*krumpli* (Hu), krum-pli
*krumplipüré* (Hu), krum-pli-pü-ré
*krumplisaláta* (Hu), krum-pli-sa-lá-ta
*krupična kaše* (Cz), kru-pič-ná ka-še
*krupnik* (Po), krup-nik
*krupuk udang* (In), kru-puk u-dang
*krusbär* (Sw), krus-bär
*kruška* (SC), kru-ška
*krustader* (Da), krus-ta-der
*krydderier* (Nw), krydd-e-ri-er
*kryddor* (Sw), kryd-dor
*kryddost* (Sw), krydd-ost
*kryddsill* (Sw), krydd-sill
*krydret and* (Da), kry-dret and
*kuah* (In), kuah
*kubis* (In), ku-bis

*Kuchen* (Gr), Ku-chen
*kuchitori* (Jp), ku-chi-to-ri
*kudamono* (Jp), ku-da-mo-no
*kue dadar* (In), ku-e da-dar
*kue kering* (In), ku-e kering
*kue lapis* (In), ku-e la-pis
*kufta* (Ar), kuf-ta
*kufta mabrouma* (Ar), kuf-ta ma-brou-ma
*kugel* (Jw), ku-gel
*Kugelhopf* (Gr-Austria), Ku-gel-hopf
*kŭ-gwā* (Ch), kŭ-gwā
*kuha* (Fi), ku-ha
*kuiken* (Du), kui-ken
*kujira* (Jp), ku-ji-ra
*kuk* (Kr), kuk
*kukkakaali* (Fi), kuk-ka-kaa-li
*kukkī* (Jp), kuk-kī
*kukurjeebhi* (Ia), ku-kur-jee-bhi
*kukurydza* (Po), ku-ku-ry-dza
*kuku wa kuchoma* (Af-Swahili), ku-ku wa ku-cho-ma
*kuku wa kukaanga* (Af-Swahili), ku-ku wa ku-ka-a-nga
*kulebiak* (Po), ku-le-biak
*kulebyaka* (Rs), ku-le-bya-ka
*kulfi* (Ia), kul-fi
*kulich* (Rs), ku-lich
*kumiss* (Rs), ku-miss
*kumle* (Nw), kum-le
*kummel* (Sw), kum-mel
*Kümmel* (Gr), Küm-mel
*kumminost* (Sw), kum-min-ost
*kumquat* (US), kum-quat

*kunde* (Af-Swahili), ku-nde
*kuneli* (Gk), ku-ne-li
*kung pao* (Th), kung pao
*kùng shui* (Ch), kùng shui
*kunsei no nishin* (Jp), kun-se-i no ni-shin
*kunyit* (In), ku-nyit
*kupus* (SC), ku-pus
*kura* (Po), ku-ra
*kurapatka* (Rs), ku-ra-pat-ka
*kura smażona* (Po), ku-ra sma-żo-na
*Kürbis* (Gr), Kür-bis
*kurcze z jarzynami* (Po), kur-cze z jar-zy-na-mi
*kuře* (Cz), ku-ře
*kuri* (Rs), ku-ri
*kuri* (Jp), ku-ri
*kurkku* (Fi), kurk-ku
*kurkkukeitto* (Fi), kurk-ku-ke-it-to
*kurkkusalaatti* (Fi), kurk-ku-sa-laat-ti
*kurma* (In), kur-ma
*kurnik* (Rs), kur-nik
*kuro pan* (Jp), ku-ro pan
*kurpitsa* (Fi), kur-pit-sa
*kuru üzüm* (Tr), ku-ru ü-züm
*kurz angebraten* (Gr), kurz an-ge-bra-ten

*kushizashi* (Jp), ku-shi-za-shi
*kuşkonmaz* (Tr), kuş-kon-maz
*kuvana jaja* (SC), ku-va-na ja-ja
*kuvano* (SC), ku-va-no
*kuzu* (Tr), ku-zu
*kuzumanju* (Jp), ku-zu-man-ju
*kuzu pirzolasi* (Tr), ku-zu pir-zo-la-si
*kvass* (Rs), kvass
*kveite* (Nw), kvei-te
*květák* (Cz), kvě-ták
*kwark* (Du), kwark
*kwaśne* (Po), kwaś-ne
*kyabetsu* (Jp), ky-a-be-tsu
*kyckling* (Sw), kyck-ling
*kycklinggryta* (Sw), kyck-ling-gry-ta
*kyckling ugnsstekt* (Sw), kyck-ling ugns-stekt
*kyljys* (Fi), kyl-jys
*kylling* (Nw), ky-lling
*kyllinger stegte* (Da), kyl-ling-er steg-te
*kyselé okurky* (Cz), ky-se-lé o-kur-ky
*kyūri* (Jp), ky-ū-ri
*kyūri no sumomi* (Jp), ky-ū-ri no su-mo-mi

# L

*laban zabadi* (Ar), la-ban za-ba-di
*labra* (Ia), lab-ra
*là-cháng* (Ch), là-cháng
*Lachs* (Gr), Lachs
*Lacrima Christi* (It), La-cri-ma Chris-ti
*lactose* (US), lac-tose
*là de* (Ch), là de
*là de jye-mwò jyàng* (Ch), là de jye-mwó jyàng
*ladyfingers* (US), la-dy-fin-gers
*là-gen* (Ch), là-gen
*lagkage* (Da), lag-ka-ge
*lagosta* (Pg), la-gos-ta
*lagostins* (Pg), la-gos-tins
*lágy tojás* (Hu), lágy to-jás
*làhana* (Tr), lâ-ha-na
*lahm* (Ar), lahm
*lahna* (Fi), lah-na
*lait* (Fr), lait
*laitue* (Fr), lai-tue
*laitue chicorée* (Fr), lai-tue chic-o-rée
*laitues braisées* (Fr), lai-tu-es brai-sées
*là jyàng* (Ch), là jyàng
*là jyau* (Ch), là jyau
*lakkoja* (Fi), lak-ko-ja

*laks* (Da, Nw), laks
*laksørred* (Da), laks-ør-red
*lam* (Nw), lam
*lamb* (US), lamb
*lambrópsomo* (Gk), lam-bró-pso-mo
*lamb's lettuce* (US), lamb's let-tuce
*lamb's quarters* (US), lamb's quar-ters
*lamm* (Sw), lamm
*lammaskaali* (Fi), lam-mas-kaa-li
*lammefrikassé* (Da), lam-me-fri-kas-sé
*lammekotelett* (Da, Nw), lam-me-ko-te-lett
*lammestek* (Nw), lam-me-stek
*Lammfleisch* (Gr), Lamm-fleisch
*lampaanliha* (Fi), lam-paan-li-ha
*lamponi* (It), lam-po-ni
*lamsvlees* (Du), lams-vlees
*Lancashire* (GB), Lanc-a-shire
*langosta* (Sp), lan-gos-ta
*langouste* (Fr), lan-gouste

*langoustine* (Fr), lan-gou-
stine
*langue* (Fr), langue
*langue de boeuf* (Fr), langue
de boeuf
*langue de boeuf gelée* (Fr),
langue de boeuf ge-lée
*langue de chat* (Fr), langue
de chat
*lanttu* (Fi), lant-tu
*lanttulaatikko* (Fi), lant-tu-
laa-tik-ko
*lapereau* (Fr), la-pe-reau
*lapin* (Fr), la-pin
*lappi* (Fi), lap-pi
*lapskaus* (Nw), laps-kaus
*laranja* (Pg), la-ran-ja
*laranjada* (Pg), la-ran-ja-da
*lard* (US), lard
*lard fumé* (Fr), lard fu-mé
*larding* (US), lard-ing
*lardon* (Fr), lar-don
*lardoon* (US), lar-doon
*lasagne* (It), la-sa-gne
*lasagne verde* (It), la-sa-gne
ver-de
*lasimestarin silli* (Fi), la-si-
mes-ta-rin sil-li
*laskiaispulla* (Fi), las-ki-ais-
pul-la
*lassi* (Ia), las-si
*lathee* (Gk), la-thee
*latke* (Jw), lat-ke
*latte* (It), lat-te
*lättstekt* (Sw), lätt-stekt
*lattuga* (It), lat-tu-ga
*Lauch* (Gr), Lauch
*làu-myàn* (Ch), làu-myàn

*lavagante* (Pg), la-va-gan-te
*laver* (US), la-ver
*lax* (Sw), lax
*lazac* (Hu), la-zac
*leather* (US), leath-er
*lebbencsleves* (Hu), leb-ben-
cs-le-ves
*leben* (Jw-Israel), le-ben
*Leber* (Gr), Le-ber
*Leberknödelsuppe* (Gr), Le-
ber-knö-del-sup-pe
*Leberwurst* (Gr), Le-ber-
wurst
*Lebkuchen* (US), Leb-kuch-
en
*leche* (Sp), le-che
*leche de manteca* (Sp), le-
che de man-te-ca
*lechem* (Jw-Israel), le-chem
*leche quemada* (Sp), le-che
que-ma-da
*lecithin* (US), lec-i-thin
*lecsó* (Hu), le-csó
*led* (Cz), led
*Lederzucker* (Gr), Le-der-
zuc-ker
*leek* (US), leek
*lefse* (Nw), lef-se
*légumes* (Fr), lé-gumes
*legumes* (Pg), le-gu-mes
*Leidse kaas* (Du), Leid-se
kaas
*leikkeleitä* (Fi), le-ik-ke-le-i-
tä
*leipä* (Fi), le-i-pä
*leitão* (Pg), le-i-tã-o
*leite* (Pg), le-i-te
*leite creme* (Pg), le-i-te creme

*lekvár* (Hu), lek-vár
*lekváros derelye* (Hu), lek-vá-ros de-re-lye
*lemon* (US), lem-on
*lemon balm* (US), lem-on balm
*lemongrass* (US), lem-on-grass
*lencseleves* (Hu), len-cse-le-ves
*lenguado* (Sp), len-gua-do
*lentejas* (Sp), len-te-jas
*lenticchi* (It), len-tic-chi
*lentilhas* (Pg), len-ti-lhas
*lentille* (Fr), len-tille
*lepre* (It), le-pre
*lesni jahody* (Cz), les-ni ja-ho-dy
*less* (US), less
*letas* (Af-Swahili), le-tas
*letterbanket* (Du), let-ter-ban-ket
*lettuce* (US), let-tuce
*leuqkuas* (In), leuq-kuas
*levadura* (Sp), le-va-du-ra
*lever* (Da, Du, Nw, Sw), le-ver
*leverpostej* (Da), le-ver-po-stej
*levrant* (Fr), le-vrant
*levrek* (Tr), lev-rek
*levure* (Fr), le-vure
*levure de bière* (Fr), le-vure de bi-ère
*liba* (Hu), li-ba
*libamájas kifli* (Hu), li-ba-má-jas kif-li
*libové* (Cz), li-bo-vé
*licorice* (US), lic-o-rice
*liebre* (Sp), li-e-bre

*Liederkranz* (US), Lie-der-kranz
*liemi* (Fi), li-e-mi
*lievito* (It), lie-vi-to
*lievito di birra* (It), lie-vi-to di bir-ra
*lièvre* (Fr), li-èvre
*light* (US), light
*lights* (GB), lights
*libajuuresmuhennos* (Fi), li-ha-juu-res-mu-hen-nos
*libakaalilaatikko* (Fi), li-ha-kaa-li-laa-tik-ko
*libakeitto* (Fi), li-ha-ke-it-to
*libakobokas* (Fi), li-ha-ko-ho-kas
*libaliemi* (Fi), li-ha-li-e-mi
*libamureke* (Fi), li-ha-mu-re-ke
*libamurekekääryleet* (Fi), li-ha-mu-re-ke-kää-ry-leet
*libamurekekakku* (Fi), li-ha-mu-re-ke-kak-ku
*libamurekepiiras* (Fi), li-ha-mu-re-ke-pii-ras
*libamurekevoileivät* (Fi), li-ha-mu-re-ke-voi-lei-vät
*libapyörykät* (Fi), li-ha-py-ö-ry-kät
*lilikoi* (Pl-Hawaii), li-li-ko-i
*lima bean* (US), li-ma bean
*Limabohnen* (Gr), Li-ma-boh-nen
*limão* (Pg), li-mā-o
*limau* (Af-Swahili), li-mau
*limba cu misline* (Ro), lim-ba cu mi-sli-ne
*Limburger* (Gr), Lim-burg-er

*limburský sýr* (Cz), lim-bur-
ský sýr
*lime* (US), lime
*limon* (Fr), li-mon
*limon* (Tr), li-mon
*limonada* (Pg), li-mo-na-da
*limonade* (Du), li-mo-na-de
*limonata* (It), li-mo-na-ta
*limone* (It), li-mo-ne
*limpa* (Sw), lim-pa
*limppukukko* (Fi), limp-pu-
kuk-ko
*limu* (Pl-Hawaii), li-mu
*lin* (Cz), lin
*lingon* (Sw), ling-on
*lingonberry* (US), ling-on-
ber-ry
*lingua* (It), lin-gua
*lingua di bue* (It), lin-gua di
bu-e
*lingua di vitello* (It), lin-gua
di vi-tel-lo
*linguado* (Pg), lin-gua-do
*linguiça* (Pg), lin-gui-ça
*linguine* (It), lin-gui-ne
*Linsensuppe* (Gr), Lin-sen-
sup-pe
*Linzertorte* (Gr-Austria), Linz-
er-torte
*lipeäkala* (Fi), li-pe-ä-ka-la
*lipid* (US), lip-id
*liptói körözött* (Hu), lip-tói
kö-rö-zött
*liqueur* (US), li-queur
*liquid sugar* (US), liq-uid
sug-ar
*liquor* (US), liq-uor
*liquore* (It), li-quo-re
*lishki* (Rs), lish-ki

*lískové ořechy* (Cz), lís-ko-vé
o-ře-chy
*list* (SC), list
*litchi* (Ch), li-tchi
*lithe* (Sc), lithe
*littleneck clam* (US), lit-tle-
neck clam
*lívance* (Cz), lí-van-ce
*Livarot* (Fr), Li-va-rot
*liver sausage* (US), li-ver sau-
sage
*lobescoves* (Da), lobes-coves
*lobster* (US), lob-ster
*lobster Newburg* (US), lob-
ster New-burg
*lobster sauce* (US), lob-ster
sauce
*lobster Thermidor* (US), lob-
ster Ther-mi-dor
*loco moco* (Pl), lo-co mo-co
*lody* (Po), lo-dy
*løg* (Da), løg
*loganberry* (US), lo-gan-ber-
ry
*lohi* (Fi), lo-hi
*lohilaatikko* (Fi), lo-hi-laa-
tik-ko
*lohipiirakka* (Fi), lo-hi-pii-
rak-ka
*lobz* (Ar), lohz
*lök* (Sw), lök
*løk* (Nw), løk
*löksoppa* (Sw), lök-sop-pa
*lokum* (Tr), lo-kum
*lombata di vitello* (It), lom-
ba-ta di vi-tel-lo
*lombo de vitela* (Pg), lom-bo
de vi-te-la
*lombok* (In), lom-bok

**lo mein** (Ch), lo mein
**lomi-lomi** (Pl-Hawaii), lo-mi-lo-mi
**lomo** (Sp), lo-mo
**London broil** (US), Lon-don broil
**longan** (Ch), lon-gan
**long-billed sturgeon** (US), long-billed stur-geon
**longe** (Fr), longe
**long-grain rice** (US), long-grain rice
**long rice** (US), long rice
**lontong** (In), lon-tong
**loquat** (US), lo-quat
**löskokta ägg** (Sw), lös-kok-ta ägg
**losos** (Cz), lo-sos
**losoś** (Po), lo-soś
**lososina zapechonnaia v fol'ge** (Rs), lo-so-si-na za-pe-chon-na-ia v fol'ge
**lotte** (Fr), lotte
**lotte vapeur** (Fr), lotte va-peur
**lotus** (US), lo-tus
**Louis sauce** (US), Lou-is sauce
**loukanika** (Gk), lou-ka-ni-ka
**loukanikopita** (Gk), lou-ka-ni-ko-pi-ta

**loukoumades** (Gk), lou-kou-ma-des
**loup** (Fr), loup
**loup de mer au fenouil** (Fr), loup de mer au fen-ouil
**loupáček** (Cz), lo-u-pá-ček
**lovage** (US), lov-age
**low-calorie** (US), low-cal-o-rie
**low-fat** (US), low-fat
**low in cholesterol** (US), low in cho-les-ter-ol
**low in saturated fat** (US), low in sat-u-rat-ed fat
**low-sodium** (US), low-sod-i-um
**low-sodium milk** (US), low-sod-i-um milk
**lox** (Jw), lox
**luk** (Rs), luk
**lula** (Pg), lu-la
**lulu frita** (Pg), lu-lu fri-ta
**lumache** (It), lu-ma-che
**lumpfish roe** (US), lump-fish roe
**luostari** (Fi), lu-os-ta-ri
**luquete** (Sp), lu-que-te
**lutefisk** (Nw), lu-te-fisk
**lutfisk** (Sw), lut-fisk
**luumu** (Fi), luu-mu
**lyonnaise, à la** (Fr), ly-on-naise, à la

# M

**maandazi** (Af-Swahili), ma-a-nda-zi

**maatjes haring** (Du), maatjes har-ing

**maçã** (Pg), ma-çã

**macadamia nut** (US), mac-a-da-mi-a nut

**macaroni** (US), mac-a-ro-ni

**macaroni and cheese** (US), mac-a-ro-ni and cheese

**macaroons** (US), mac-a-roons

**macarrao** (Pg), ma-ca-rrao

**macarrones** (Sp), ma-ca-rro-nes

**maccheroni** (It), mac-che-ro-ni

**mace** (US), mace

**macédoine** (Fr), ma-cé-doine

**macedonia di frutta** (It), ma-ce-do-ni-a di frut-ta

**mâche** (Fr), mâche

**mackerel** (US), mack-er-el

**mackerel shark** (US), mack-er-el shark

**Madeira** (Pg), Ma-dei-ra

**madeleine** (Fr), ma-de-leine

**Maderawein** (Gr), Ma-der-a-wein

**madère, sauce au** (Fr), ma-dère, sauce au

**madrilène** (Fr), ma-dri-lène

**magert** (Sw), ma-gert

**maggiorana** (It), mag-gio-ra-na

**magnesium** (US), mag-nes-i-um

**magro, di** (It), ma-gro, di

**maguro** (Jp), ma-gu-ro

**maguro no sashimi** (Jp), ma-gu-ro no sa-shi-mi

**mahi-mahi** (Pl), ma-hi-ma-hi

**mahlepi** (Ar), mah-le-pi

**ma ho** (Th), ma ho

**mai** (Ar), mai

**maiale** (It), ma-ia-le

**maialino di latte** (It), ma-ia-li-no di lat-te

**maida** (Ia), mai-da

**mài-dz** (Ch), mài-dz

**Maifisch** (Gr), Mai-fisch

**maigre, au** (Fr), mai-gre, au

**maini** (Af-Swahili), ma-i-ni

**maionese** (It, Pg), ma-io-ne-se

**mài-pyàn** (Ch), mài-pyàn

**mais** (Fr), ma-is

**mais** (Nw), ma-is

**Mais** (Gr), Mais

**maisild** (Nw), mai-sild
**Maismehl** (Gr), Mais-mehl
**maison** (Fr), mai-son
**maito** (Fi), ma-i-to
**maitre d'hotel beurre** (Fr),
maitre d'ho-tel beurre
**maíz** (Sp), ma-íz
**maize** (US), maize
**máj** (Hu), máj
**májas hurka** (Hu), má-jas
hur-ka
**majfisk** (Sw), maj-fisk
**májgombócleves** (Hu), máj-
gom-bóc-le-ves
**maji** (Af-Swahili), ma-ji
**maji ya dafu** (Af-Swahili),
ma-ji ya da-fu
**maji ya matunda** (Af-Swahili), ma-ji ya ma-tu-nda
**majoneesi** (Fi), ma-jo-nee-si
**majorana** (Sp), ma-jo-ra-na
**májpastetom** (Hu), máj-pas-
te-tom
**majsild** (Da), maj-sild
**makarna** (Tr), ma-kar-na
**makaron** (Po), ma-ka-ron
**makaroni** (Fi), ma-ka-ro-ni
**makaronia me kima** (Gk),
ma-ka-ron-ia me ki-ma
**makarunen** (Jw), ma-ka-ru-
nen
**makimaki** (Jp), ma-ki-ma-ki
**makizushi** (Jp), ma-ki-zu-shi
**makkara** (Fi), mak-ka-ra
**makkhani murgaa** (Ia),
mak-kha-ni mur-gaa
**mako** (US), ma-ko
**mákos kalács** (Hu), má-kos
ka-lács

**mákos metélt** (Hu), má-kos
me-télt
**makowiec** (Po), ma-ko-wiec
**makreel** (Du), ma-kreel
**makrel** (Da), ma-krel
**Makrele** (Gr), Ma-kre-le
**makrell** (Nw), ma-krell
**makrill** (Sw), ma-kriil
**makrilli** (Fi), mak-ril-li
**makrut** (Th), mak-rut
**maksa** (Fi), mak-sa
**maksalaatikko** (Fi), mak-sa-
laa-tik-ko
**malacpecsenye** (Hu), ma-
lac-pe-cse-nye
**malasol** (Rs), ma-la-sol
**maline** (SC), ma-li-ne
**mǎ-líng-shǔ** (Ch), mǎ-líng-
shǔ
**mǎ-líng-shǔ shā-là** (Ch), mǎ-
líng-shǔ shā-là
**maliny** (Cz, Po), ma-li-ny
**málna** (Hu), mál-na
**malt** (US), malt
**maltaise, à la** (Fr), mal-taise,
à la
**maltaise, sauce** (Fr), mal-
taise, sauce
**malt bread** (GB), malt bread
**malt extract** (US), malt ex-
tract
**mămăligá** (Ro), mă-mă-li-gá
**mamao** (Pg), ma-mao
**mame** (Jp), ma-me
**mämmi** (Fi), mäm-mi
**mandariini** (Fi), man-da-rii-
ni
**mandarinka** (Cz), man-da-
rin-ka

*mandarino* (It), man-da-ri-no

*mandarin orange* (US), man-dar-in or-ange

*mandelbiskvier* (Sw), man-del-bisk-vi-er

*Mandelbrezeln* (Gr), Man-del-bre-zeln

*Mandeln* (Gr), Man-deln

*Mandeltorte* (Gr), Man-del-tor-te

*mandlar* (Sw), mand-lar

*mandler* (Nw), mand-ler

*mandorla* (It), man-dor-la

*mandorla amara* (It), man-dor-la a-ma-ra

*mandorle tostate* (It), man-dor-le tos-ta-te

*manestra* (Gk), ma-nes-tra

*manga* (Pg), man-ga

*manganese* (US), man-ga-nese

*mange-tout* (Fr), mange-tout

*máng-gwo* (Ch), máng-gwo

*mango* (US), man-go

*mango chutney* (Ia), man-go chut-ney

*Mangold* (Gr), Man-gold

*mangosteen* (US), man-go-steen

*Manhattan clam chowder* (US), Man-hat-tan clam chow-der

*manicotti* (It), ma-ni-cot-ti

*manioc* (Fr), man-i-oc

*manitaria* (Gk), ma-ni-tar-ia

*manjarblanco* (Sp), man-jar-blan-co

*manju* (Jp), man-ju

*mannagrynspudding* (Sw), man-na-gryns-pud-ding

*mannapuuro* (Fi), man-na-puu-ro

*mansikkalumi* (Fi), man-sik-ka-lu-mi

*mansikoita* (Fi), man-si-ko-i-ta

*mantar* (Tr), man-tar

*manteiga* (Pg), man-te-i-ga

*mantequilla* (Sp), man-te-qui-lla

*mán-tóu* (Ch), mán-tóu

*manty* (Rs), man-ty

*màn-yú* (Ch), màn-yú

*manzana* (Sp), man-za-na

*manzo* (It), man-zo

*manzo arrosto* (It), man-zo ar-ro-sto

*manzo brasiato* (It), man-zo bra-sia-to

*manzo salato* (It), man-zo sa-la-to

*maple syrup* (US), ma-ple syr-up

*maquereaux* (Fr), ma-que-reaux

*maquereaux au vin blanc* (Fr), ma-que-reaux au vin blanc

*maquereaux marine* (Fr), ma-que-reaux ma-ri-ne

*marak* (Jw-Israel), ma-rak

*marak perot kar* (Jw-Israel), ma-rak pe-rot kar

*marak yerakot* (Jw-Israel), ma-rak ye-ra-kot

*maräng* (Sw), ma-räng

*Maraschino* (It), Mar-a-schi-no

**marchands de vin, sauce** (Fr), mar-chands de vin, sauce

**marchewka** (Po), mar-chew-ka

**marchewka w sosie** (Po), mar-chew-ka w so-sie

**marelica** (SC), ma-re-li-ca

**margarine** (US), mar-ga-rine

**marguery** (Fr), mar-gue-ry

**marhaeröleves** (Hu), mar-ha-e-rö-le-ves

**marhahús** (Hu), mar-ha-hús

**marhasült** (Hu), mar-ha-sült

**maridhes** (Gk), mar-i-dhes

**Marienkraut** (Gr), Ma-ri-en-kraut

**marigold** (US), mar-i-gold

**Marille** (Gr-Austria), Ma-ril-le

**Marillenknödel** (Gr-Austria), Ma-ril-len-knö-del

**marinade** (US), mar-i-nade

**marinara, alla** (It), ma-ri-na-ra, al-la

**marinata** (It), mar-i-na-ta

**marinate** (US), mar-i-nate

**mariné** (Fr), ma-ri-né

**marinière, à la** (Fr), ma-ri-nière, à la

**marinovannye griby** (Rs), ma-ri-no-van-nye gri-by

**marinvota silke** (Rs), ma-rin-vo-ta sil-ke

**mariscos** (Sp), mar-is-cos

**maritozzi** (It), ma-ri-toz-zi

**marjolaine** (Fr), mar-jo-laine

**marjoram** (US), mar-jo-ram

**markjordbær** (Nw), mark-jord-bær

**Markklösschen** (Gr), Mark-klös-schen

**marmalade** (US), mar-ma-lade

**marmellata** (It), mar-mel-la-ta

**marmellata di cotogne** (It), mar-mel-la-ta di co-to-gne

**marmelo** (Pg), mar-me-lo

**maroilles** (Fr), mar-oi-lles

**marron** (Fr), mar-ron

**marrons glacé** (Fr), mar-rons gla-cé

**marrow (US) mar-row**

**Marsala** (It), Mar-sa-la

**marshmallow** (US), marsh-mal-low

**marsh rabbit** (US), marsh rab-bit

**mártás** (Hu), már-tás

**marul** (Tr), ma-rul

**Maryland fried chicken** (US), Mar-y-land fried chick-en

**marynowane sledzie** (Po), ma-ry-no-wa-ne sle-dzie

**marzipan** (Fr), mar-zi-pan

**masa** (Mx), ma-sa

**masa harina** (Mx), ma-sa ha-ri-na

**masala** (Ia), ma-sa-la

**mascarpone** (It), mas-car-po-ne

**mascotte, à la** (Fr), mas-cotte, à la

**masgoof** (Ar), mas-goof

**mash** (US), mash

**mash, to** (US), mash, to

**masline** (SC), ma-sli-ne

**maslo** (Po), mas-lo

**máslo** (Cz), más-lo
**maso** (Cz), ma-so
**masové knedlíčky** (Cz), ma-so-vé kned-líč-ky
**maso z divokého kance** (Cz), ma-so z d-i-vo-ké-ho kan-ce
**Masséna, à la** (Fr), Mas-sén-a, à la
**masu** (Jp), ma-su
**masur** (Ia), mas-ur
**mat** (Nw), mat
**matcha** (Jp), mat-cha
**maté** (Sp), ma-té
**matelote** (Fr), mat-e-lote
**matfett** (Nw), mat-fett
**má-tí** (Ch), má-tí
**mäti ja paahtoleipä** (Fi), mä-ti ja paah-to-le-i-pä
**Matjeshering in saurer Sahne** (Gr), Mat-jes-her-ing in sau-rer Sah-ne
**matjessill** (Sw), mat-jes-sill
**matsutake** (Jp), ma-tsu-ta-ke
**matzo** (Jw), mat-zo
**matzo balls** (Jw), mat-zo balls
**matzo brei** (Jw), mat-zo brei
**matzo meal** (Jw), mat-zo meal
**Maultaschen** (Gr), Maul-ta-schen
**mausteita** (Fi), ma-us-te-i-ta
**mayai ya kuchemsha yaliyo laini** (Af-Swahili), ma-ya-i ya ku-che-msha ya-li-yo la-i-ni
**mayai ya kuchemsha yaliyo magumu** (Af-Swahili), ma-ya-i ya ku-che-msha ya-li-yo ma-gu-mu

**mayai ya kuvuruga** (Af-Swahili), ma-ya-i ya ku-vu-ru-ga
**mayeritsa** (Gk), ma-ye-rit-sa
**mayim** (Jw-Israel), ma-yim
**mayonnaise** (Fr), may-on-naise
**mazarintårta** (Sw), ma-za-rin-tår-ta
**maziwa** (Af-Swahili), ma-zi-wa
**maziwa ya kuganda** (Af-Swahili), ma-zi-wa ya ku-ga-nda
**mazorka de maíz** (Mx), ma-zor-ka de ma-íz
**mazurek** (Po), ma-zu-rek
**mazurka** (Rs), ma-zur-ka
**mbaazi** (Af-Swahili), m-ba-a-zi
**mchele** (Af-Swahili), m-che-le
**mchicha** (Af-Swahili), m-chi-cha
**mchicha wa nazi** (Af-Swahili), m-chi-cha wa na-zi
**mchuzi** (Af-Swahili), m-chu-zi
**mchuzi wa kuku** (Af-Swahili), m-chu-zi wa ku-ku
**mchuzi wa nyama** (Af-Swahili), m-chu-zi wa n-ya-ma
**mdzhavai kombosto** (Rs), mdzha-vai kom-bos-to
**mead** (US), mead
**mealiepap** (Af), meal-ie-pap
**meat** (US), meat
**meat extender** (US), meat ex-ten-der
**mečoun** (Cz), me-čo-un

**médaillon** (Fr), mé-dai-llon
**médaillons de veau** (Fr), mé-dai-llons de veau
**medamayaki** (Jp), me-da-ma-ya-ki
**medio crudo** (Sp), me-di-o cru-do
**medisterkaker** (Nw), me-dis-ter-ka-ker
**medivnyk** (Rs), me-div-nyk
**medlar** (US) med-lar
**mee krob** (Th), mee krob
**Meeresfrüchte** (Gr), Mee-res-früch-te
**Meerrettich** (Gr), Meer-ret-tich
**Meerrettichsosse** (Gr), Meer-ret-tich-sos-se
**meggy** (Hu), meggy
**meggyes rétes** (Hu), meg-gyes ré-tes
**Mehlspeisen** (Gr), Mehl-spei-sen
**mebshi** (Ar), meh-shi
**mebu** (Fi), me-hu
**mei** (Pl), me-i
**méi-dz** (Ch), méi-dz
**mejillónes** (Sp), me-ji-lló-nes
**mel** (Nw), mel
**mela** (It), me-la
**melagrana** (It), me-la-gra-na
**melancia** (Pg), me-lan-ci-a
**mélangée** (Fr), mé-lan-gée
**melanzane** (It), me-lan-za-ne
**Melanzani** (Gr-Austria), Me-lan-za-ni
**melão** (Pg), me-lã-o
**Melba toast** (US), Mel-ba toast
**melcocha** (Sp), mel-co-cha

**mele fritte al rum** (It), me-le frit-te al rum
**melidzanes** (Gk), me-lid-za-nes
**melidzanosalata** (Gk), me-lid-za-no-sa-la-ta
**melk** (Du, Nw), melk
**melkbrood** (Du), mel-k-brood
**melktert** (Af), melk-tert
**meloa** (Pg), me-lo-a
**melocotón** (Sp), me-lo-co-tón
**meloen** (Du), me-loen
**melokhia** (Ar-Egypt), me-lo-khia
**melon** (US), mel-on
**melon** (Fr, Nw, Sw), me-lon
**melón** (Sp), me-lón
**Melone** (Gr), Me-lo-ne
**melone e prosciutto** (It), me-lo-ne e pro-sciut-to
**meloun** (Cz), me-lo-un
**melt, to** (US), melt, to
**menestra** (Sp), me-nes-tra
**menrui** (Jp), men-ru-i
**menta** (It), men-ta
**mentega** (Ml), men-te-ga
**menthe** (Fr), menthe
**menthe poivrée** (Fr), menthe poi-vrée
**meriantura** (Fi), me-ri-an-tu-ra
**meringue glacée** (Fr), me-ringue gla-cée
**merlan** (Fr), mer-lan
**merluzzo** (It), mer-luz-zo
**mermelada** (Sp), mer-me-la-da
**meruňky** (Cz), me-ruň-ky

**meshimono** (Jp), me-shi-mo-no

**mesimarja** (Fi), me-si-mar-ja

**meso od divljači** (SC), me-so od div-lja-či

**metélt** (Hu), me-télt

**metsäsieniä** (Fi), met-sä-si-e-ni-ä

**metso** (Fi), met-so

**Mettwurst** (Gr), Mett-wurst

**meunière** (Fr), meu-nière

**Mexican tea** (US), Mex-i-can tea

**mexilhãos** (Pg), me-xi-lhã-os

**meyve** (Tr), mey-ve

**méz** (Hu), méz

**mezedhakia** (Gk), me-ze-dha-kia

**mézeskalács** (Hu), mé-zes-ka-lács

**mezza** (Ar), mez-za

**mezza bishurba** (Ar), mez-za bi-shur-ba

**miasnaia solianka** (Rs), mias-na-ia so-lian-ka

**míchaná vejce** (Cz), mí-cha-ná vej-ce

**microwave oven** (US), mi-cro-wave o-ven

**midhia** (Gk), mi-dhia

**midya** (Tr), mid-ya

**midya dolmasi** (Tr), mid-ya dol-ma-si

**mie goreng** (In), mie go-reng

**miel** (Fr), miel

**miele** (It), mi-e-le

**mie pangsit** (In), mie pang-sit

**mięso** (Po), mię-so

**mì-gān** (Ch), mì-gān

**mihallabiyya** (Ar), mi-hal-la-biy-ya

**mijoté** (Fr), mi-jo-té

**mí jyou** (Ch), mí jyou

**mikan** (Jp), mi-kan

**milanese, alla** (It), mi-la-ne-se, al-la

**Milch** (Gr), Milch

**milho** (Pg), mi-lho

**milk** (US), milk

**milk powder** (US), milk pow-der

**milk substitutes** (US), milk sub-sti-tutes

**milk, vegetable** (US), milk, veg-e-ta-ble

**mille-fanti** (It), mil-le-fan-ti

**mille-feuille** (Fr), mille-feu-ille

**millet** (US), mil-let

**milt** (US), milt

**mimosa** (Fr), mi-mo-sa

**mince, to** (US), mince, to

**mincemeat** (US), mince-meat

**mince pie** (US), mince pie

**minerálka** (Cz), mi-ne-rál-ka

**mineralvann** (Nw), mi-ne-ral-vann

**mineralvatten** (Sw), mi-ne-ral-vat-ten

**Mineralwasser** (Gr), Mi-ne-ral-was-ser

**minestra di verdure** (It), mi-nes-tra di ver-du-re

**minestrina** (It), min-e-stri-na

**minestrone** (It), min-e-stro-ne

**minestrone alla genovese** (It), min-e-stro-ne al-la ge-no-ve-se

**mint** (US), mint
**Minze** (Gr), Min-ze
**miodowo-orzechowy ma-zurek** (Po), mio-do-wo-or-ze-chowy ma-zu-rek
**miolos** (Pg), mi-o-los
**mirabeau** (Fr), mi-ra-beau
**mirabelle** (Fr), mir-a-belle
**mirepoix** (Fr), mire-poix
**mirin** (Jp), mi-rin
**miringbe** (It), mi-rin-ghe
**mirliton** (US), mir-li-ton
**miroton** (Fr), mi-ro-ton
**mirtillo** (It), mir-til-lo
**mirugai** (Jp), mi-ru-ga-i
**mì rwan-ji** (Ch), mì rwan-ji
**mishmishi** (Af-Swahili), mish-mish-i
**misir** (Tr), mi-sir
**miso** (Jp), mi-so
**misoshiru** (Jp), mi-so-shi-ru
**misto** (It), mis-to
**misto alla griglia** (It), mis-to al-la grig-lia
**misto bosco** (It), mis-to bos-co
**misto di verdure** (It), mis-to di ver-du-re
**mitili** (It), mi-ti-li
**mititei** (Ru), mi-ti-tei
**mitsuba** (Jp), mi-tsu-ba
**mitsumame** (Jp), mi-tsu-ma-me
**mitzutaki** (Jp), mi-tzu-ta-ki
**mix, to** (US), mix, to
**mixte** (Fr), mixte
**mizu** (Jp), mi-zu
**mizutake** (Jp), mi-zu-ta-ke
**mjölk** (Sw), mjölk

**mjuka småfranska** (Sw), mju-ka små-fran-ska
**mjukost** (Sw), mjuk-ost
**mkate na siagi** (Af-Swahili), m-ka-te na si-a-gi
**mleczko cielece potrawie** (Po), mlecz-ko cie-le-ce po-tra-wie
**mleko** (Cz, Po), mle-ko
**mleté hovězí** (Cz), mle-té ho-vě-zí
**mlevena govedina** (SC), mle-ve-na go-ve-di-na
**moana** (Pl), mo-a-na
**mocha** (US), mo-cha
**mochi** (Jp), mo-chi
**mochi gome** (Jp), mo-chi go-me
**mochiko** (Jp), mo-chi-ko
**mochonye** (Rs), mo-cho-nye
**mochonye arbuzy** (Rs), mo-cho-nye ar-buzy
**mock duck** (US), mock duck
**mock pork** (US), mock pork
**mock turtle soup** (US), mock tur-tle soup
**moelle** (Fr), moelle
**mofongo** (Sp), mo-fon-go
**mo gwa** (Ch), mo gwa
**Möhren** (Gr), Möh-ren
**moído** (Pg), mo-í-do
**moje** (Sp), mo-je
**moje de ajo** (Sp), mo-je de a-jo
**mokoto** (Af), mo-ko-to
**molasses** (US), mo-las-ses
**mold** (US), mold
**mole** (Mx), mo-le
**mole poblano de guajolote**

(MX), mo-le po-bla-no de gua-jo-lo-te

**mole verde** (Sp), mo-le ver-de

**molho** (Pg), mo-lho

**molho de salada** (Pg), mo-lho de sa-la-da

**molho de tomate** (Pg), mo-lho de to-ma-te

**Molke** (Gr), Mol-ke

**mollet** (Fr), mol-let

**molletes** (Sp), mol-le-tes

**mollusk** (US), mol-lusk

**molochnyi sup s risom** (Rs), mo-loch-nyi sup s ri-som

**moloko** (Rs), mo-lo-ko

**molusco** (Sp), mo-lus-co

**molusque** (Fr), mo-lusque

**momo** (Jp), mo-mo

**mondel** (Jw), mond-el

**monkey bread** (Af), mon-key bread

**monkfish** (US), monk-fish

**monosodium glutamate** (US), mon-o-so-di-um glu-ta-mate

**mont blanc** (Fr), mont blanc

**monte bianco** (It), mon-te bian-co

**Monterey Jack** (US), Mon-te-rey Jack

**montmorency, à la** (Fr), mont-mo-ren-cy, à la

**Moosbeere** (Gr), Moos-bee-re

**moqueca** (Pg), mo-que-ca

**morangos** (Pg), mo-ran-gos

**morcella** (Pg), mor-cel-la

**morel** (US), mo-rel

**morello cherry** (US), mo-rel-lo cher-ry

**moriawase** (Jp), mo-ri-a-wa-se

**morille** (Fr), mo-rille

**mornay, sauce** (Fr), mor-nay, sauce

**moromi miso** (Jp), mo-ro-mi mi-so

**morötter** (Sw), mo-röt-ter

**mořské mušle** (Cz), moř-ské mušle

**morskoi yazyk** (Rs), mor-skoi ya-zyk

**mořský okoun** (Cz), moř-ský o-ko-un

**mořští mlži** (Cz), moř-ští ml-ži

**mořští ráčci** (Cz), moř-ští ráč-ci

**mortadela** (Pg, Sp), mor-ta-de-la

**mortadella** (It), mor-ta-del-la

**mortella** (It), mor-tel-la

**mortella di palude** (It), mor-tel-la di pa-lu-de

**morue** (Fr), mo-rue

**mosselen** (Du), mos-se-len

**mostaccioli** (It), mo-stac-cio-li

**mostarda** (It, Pg), mos-tar-da

**mostarda di frutta** (It), mos-tar-da di frut-ta

**mostaza** (Sp), mos-ta-za

**mosterd** (Du), mos-terd

**moučníky** (Cz), mo-uč-ní-ky

**moules** (Fr), moules

**moules à la marinière** (Fr), moules à la ma-ri-niè-re

**moules farcies** (Fr), moules far-cies

*moulokhiya* (Ar), mou-lo-khi-ya

*mount, to* (US), mount, to

*mountain cheese* (US), moun-tain cheese

*mountain oysters* (US), moun-tain oys-ters

*moussaka* (Gk), mous-sa-ka

*mousse* (Fr), mousse

*mousse de volaille* (Fr), mousse de vo-laille

*mousseline* (Fr), mousse-line

*mousseline, sauce* (Fr), mousse-line, sauce

*mousseuse, sauce* (Fr), mous-seuse, sauce

*moutarde* (Fr), mou-tarde

*moutarde au poivre verte* (Fr), mou-tarde au poi-vre verte

*moyashi* (Jp), mo-ya-shi

*mozeček* (Cz), mo-ze-ček

*mozzarella* (It), moz-za-rel-la

*mrkev* (Cz), mr-kev

*mrkve* (SC), mrk-ve

*mshikaki* (Af-Swahili), m-shi-ka-ki

*mtama* (Af-Swahili), m-ta-ma

*mtindi* (Af-Swahili), m-ti-ndi

*mtori* (Af-Swahili), m-to-ri

*muffin* (US), muf-fin

*muhindi* (Af-Swahili), mu-hi-ndi

*muhogo* (Af-Swahili), mu-ho-go

*muhogo tamu* (Af-Swahili), mu-ho-go ta-mu

*muisjes* (Du), muis-jes

*muito condimentado* (Pg), mu-i-to con-di-men-ta-do

*mù lì chá* (Ch), mù lì chá

*mulligatawny soup* (Ia), mul-li-ga-taw-ny soup

*multer* (Nw), mul-ter

*muna ja pekoni* (Fi), mu-na ja pe-ko-ni

*munakas* (Fi), mu-na-kas

*Münchner Prinzregententorte* (Gr), Münch-ner Prinz-re-gen-ten-tor-te

*mung beans* (US), mung beans

*munkar* (Sw), mun-kar

*munuaiset* (Fi), mu-nu-a-i-set

*mûre* (Fr), mûre

*murgee* (Ia), mur-gee

*murgee do pyaza* (Ia), mur-gee do pyaz-a

*murg moghlai* (Ia), murg mogh-lai

*murg musallam* (Ia), murg mu-sal-lam

*murlins* (Ir), mur-lins

*muroja* (Fi), mu-ro-ja

*musaka od plavih patlidžana* (SC), mu-sa-ka od pla-vih pat-li-dža-na

*Muscheln* (Gr), Musch-eln

*muscoli* (It), mus-co-li

*mushi awabi* (Jp), mu-shi a-wa-bi

*mushigashi* (Jp), mu-shi-ga-shi

*mushimono* (Jp), mu-shi-mo-no

*mushita* (Jp), mu-shi-ta

*mushroom* (US), mush-room

***Muskatblüte*** (Gr), Mus-kat-blü-te
***muskmelon*** (US), musk-mel-on
***mussel*** (US), mus-sel
***musslor*** (Sw), muss-lor
***musta leipa*** (Fi), mus-ta le-i-pa
***mustár*** (Hu), mus-tár
***mustard*** (US), mus-tard
***mustarda*** (Pg), mus-tar-da
***mustard greens*** (US), mus-tard greens
***mustard oil*** (Ia), mus-tard oil
***mustikkapiirakka*** (Fi), mus-tik-ka-pii-rak-ka

***mustikoita*** (Fi), mus-ti-ko-i-ta
***mù-syū ròu*** (Ch), mù-syū ròu
***musztarda*** (Po), musz-tar-da
***mutton*** (US), mut-ton
***mutton fish*** (Aa), mut-ton fish
***muz*** (Tr), muz
***myàn-bāu*** (Ch), myàn-bāu
***myàn-bāu jywǎn*** (Ch), myàn-bāu jywǎn
***myàn-tyáu*** (Ch), myàn-tyáu
***mylta med grädde*** (Sw), myl-ta med gräd-de
***myrte*** (Fr), myrte
***myrtille*** (Fr), myr-tille
***myrtle*** (US), myr-tle

# N

***naan*** (Ia), naan
***nabemono*** (Jp), na-be-mo-no
***nabeyaki udon*** (Jp), na-be-ya-ki u-don
***nabos*** (Pg), na-bos
***Nach Art des Hauses*** (Gr), Nach Art des Hau-ses
***nachinka*** (Rs), na-chin-ka
***nachinka iz gribov dlia zraz*** (Rs), na-chin-ka iz gri-bov dlia zraz
***nachinka iz kapusty*** (Rs), na-chin-ka iz ka-pus-ty
***nachinka iz luka dlia zraz*** (Rs), na-chin-ka iz lu-ka dlia zraz
***nachinka iz miasa*** (Rs), na-chin-ka iz mia-sa
***nachos*** (Sp), na-chos
***Nachspeisen*** (Gr), Nach-spei-sen
***nadívané brambory*** (Cz), na-dí-va-né bram-bo-ry
***naeng myung*** (Kr), naeng myung
***naganegi*** (Jp), na-ga-ne-gi
***nagasari*** (In), na-ga-sa-ri
***nage, à la*** (Fr), nage, à la

**nahkiainen** (Fi), nah-ki-a-i-nen

**nakki** (Fi), nak-ki

**näkkileipä** (Fi), näk-ki-le-i-pä

**nákyp** (Cz), ná-kyp

**nalesniki** (Po), na-les-ni-ki

**nam** (Th), nam

**namako** (Jp), na-ma-ko

**namamono ni shita** (Jp), na-ma-mo-no ni shi-ta

**nama udon** (Jp), na-ma u-don

**nama yuba** (Jp), na-ma yu-ba

**nameko** (Jp), nam-e-ko

**nam pla** (Th), nam pla

**nam prik num** (Th), nam prik num

**nam prik ong** (Th), nam prik ong

**nanakusa** (Jp), na-na-ku-sa

**nanas goreng** (In), na-nas go-reng

**nanasi** (Af-Swahili), na-na-si

**nangka** (In), nang-ka

**nán-gwá** (Ch), nán-gwá

**nantaise, à la** (Fr), nan-taise, à la

**nantua, sauce** (Fr), nan-tu-a, sauce

**napa cabbage** (US), nap-a cab-bage

**Napfkuchen** (Gr), Napf-ku-chen

**naphal** (Hu), nap-hal

**Naples medlar** (US), Na-ples med-lar

**napój owocowy** (Po), na-pó-j o-wo-co-wy

**napój wyskokowy** (Po), na-pó-j wys-ko-ko-wy

**Napoleon** (Fr), Na-po-le-on

**Napoléon** (Fr), Na-po-lé-on

**napolitaine, à la** (Fr), na-po-li-taine, à la

**narancs** (Hu), na-rancs

**narancsiz** (Hu), na-rancs-iz

**naranja** (Sp), na-ran-ja

**naranja agria** (Sp), na-ran-ja a-gri-a

**narazuke** (Jp), na-ra-zu-ke

**nargesi kofta** (Ia), nar-ge-si kof-ta

**naruto** (Jp), na-ru-to

**naryal** (Ia), nar-y-al

**nashi** (Jp), na-shi

**nasi** (In), na-si

**nasi goreng** (In), na-si gor-eng

**nasi guri** (In), na-si gu-ri

**nasi kebuli** (In), na-si ke-bu-li

**nasi kuning** (In), na-si ku-ning

**nastoiki** (Rs), na-stoi-ki

**nasturtium** (US), nas-tur-tium

**nasu** (Jp), na-su

**nasu hasami age** (Jp), na-su ha-sa-mi age

**nata** (Sp), na-ta

**natsumikan** (Jp), nat-su-mi-kan

**natto** (Jp), nat-to

**Natur** (Gr), Na-tur

**Natur Schnitzel** (Gr), Na-tur Schnit-zel

**nature** (Fr), na-ture

**na'ud** (Ar), na'ud

**naudanliba** (Fi), na-u-dan-li-ha

**nău-dz** (Ch), nău-dz

**nauris** (Fi), na-u-ris

**naurisraaste** (Fi), na-u-ris-raas-te

**navarin** (Fr), na-va-rin

**navel orange** (US), na-vel or-ange

**navet** (Fr), na-vet

**navets à la bordelaise** (Fr), na-vets à la bor-de-laise

**navets au jambon** (Fr), na-vets au jam-bon

**nazi** (Af-Swahili), na-zi

**ndimu** (Af-Swahili), n-di-mu

**ndizi** (Af-Swahili), n-di-zi

**ndizi mbivu** (Af-Swahili), n-di-zi m-bi-vu

**neat's foot jelly** (GB), neat's foot jel-ly

**nectarine** (US), nec-tar-ine

**nedlagt** (Nw), ned-lagt

**nefra** (Gk), ne-fra

**negi** (Jp), ne-gi

**negimaki** (Jp), ne-gi-ma-ki

**neige** (Fr), neige

**Nelke** (Gr), Nel-ke

**nem nuong** (Vt), nem nuong

**Nemours** (Fr), Ne-mours

**nenas** (In), ne-nas

**neper** (Nw), ne-per

**nero** (It), ne-ro

**neroli** (Fr), ne-ro-li

**nerone, alla** (It), ne-ro-ne, al-la

**neslesuppe** (Nw), nes-le-sup-pe

**nespola** (It), nes-po-la

**Nesselrode** (Fr), Nes-sel-rode

**Neufchâtel** (Fr), Neuf-châ-tel

**Newburg sauce** (US), New-burg sauce

**New England boiled dinner** (US), New Eng-land boiled din-ner

**New England clam chowder** (US), New Eng-land clam chow-der

**New Jersey tea** (US), New Jer-sey tea

**nezhinskie ogurchiki** (Rs), ne-zhin-skie o-gur-chi-ki

**nguru** (Af-Swahili), n-gu-ru

**niacin** (US), ni-a-cin

**niacinamide** (US), ni-a-cin-a-mide

**niboshi** (Jp), ni-bo-shi

**niçoise, à la** (Fr), ni-çoise, à la

**niçoise salad** (US), ni-çoise sal-ad

**nicotinamide** (US), nic-o-tin-a-mide

**nid** (Fr), nid

**Niedernaver Kartoffeln** (Gr), Nie-der-na-ver Kar-tof-feln

**nierbroodje** (Du), nier-brood-je

**Nieren** (Gr), Nier-en

**Nierenbraten** (Gr), Nie-ren-bra-ten

**Nierenknödel** (Gr), Nie-ren-knö-del

**Nierenstück** (Gr), Nie-ren-stück

**niet te gaar** (Du), niet te gaar

**nigella** (US), ni-gel-la

*nigirizushi* (Jp), ni-gi-ri-zu-shi

*nijimasu no karage* (Jp), ni-ji-ma-su no ka-ra-ge

*nikomi* (Jp), ni-ko-mi

*niku* (Jp), ni-ku

*nikuan udon* (Jp), ni-ku-an u-don

*niku-dofu* (Jp), ni-ku-do-fu

*niku ryōri* (Jp), ni-ku ry-ō-ri

*nimame* (Jp), ni-ma-me

*nimbu achar* (Ia), nim-bu a-char

*nimbu chatni* (Ia), nim-bu chat-ni

*nimbu ka chaval* (Ia), nim-bu ka cha-val

*nimki* (Ia), nim-ki

*nimono* (Jp), ni-mo-no

*níng-méng* (Ch), níng-méng

*ninjin* (Jp), nin-jin

*ninniku* (Jp), nin-ni-ku

*Niolo* (Fr), Ni-o-lo

*nira* (Jp), ni-ra

*nishiki tamago* (Jp), ni-shi-ki ta-ma-go

*nishime* (Jp), ni-shi-me

*nishin* (Jp), ni-shin

*nishin no kunsei* (Jp), ni-shin no kun-sei

*nita* (Jp), ni-ta

*nitamono* (Jp), ni-ta-mo-no

*nitrates* (US), ni-trates

*nitrites* (US), ni-trites

*nitrogen* (US), ni-tro-gen

*nitsuke* (Jp), nit-su-ke

*nivernaise, à la* (Fr), ni-ver-naise, à la

*niyakko* (Jp), ni-yak-ko

*nizakana* (Jp), ni-za-ka-na

*njure* (Sw), nju-re

*noce di cocco* (It), no-ce di coc-co

*noce di vitello* (It), no-ce di vi-tel-lo

*noci* (It), no-ci

*noci del Brasile* (It), no-ci del Bras-i-le

*Nockerln* (Gr), Nock-erln

*nødder* (Da), nød-der

*nogada* (Mx), no-ga-da

*noisettes* (Fr), noi-settes

*noisettes d'agneau* (Fr), noi-settes d'agn-eau

*noisettes de chevreuil* (Fr), noi-settes de chev-reuil

*noix* (Fr), noix

*noix d'acajou* (Fr), noix d'a-ca-jou

*noix de coco* (Fr), noix de co-co

*noix du Brésil* (Fr), noix du Bré-sil

*noix muscade* (Fr), noix mus-cade

*nokedli* (Hu), nok-ed-li

*nøkkelost* (Nw), nøk-kel-ost

*nomimono* (Jp), no-mi-mo-no

*Nonnenfurz* (Gr), Non-nen-furz

*nonnette* (Fr), non-nette

*nonnutritive sweetener* (US), non-nu-tri-tive sweet-en-er

*nonpareille* (Fr), non-pa-reille

*nonvintage* (US), non-vin-tage

*nonya* (Ml), no-nya

**noodles** (US), noo-dles
**nopales** (Mx), no-pa-les
**nopales con chile pasilla** (Mx), no-pa-les con chil-e pa-si-lla
**nopales con queso** (Mx), no-pa-les con que-so
**Nordseekrabbencocktail** (Gr), Nord-see-krab-ben-cock-tail
**nori** (Jp), no-ri
**nori chazuke** (Jp), no-ri cha-zu-ke
**norimaki** (Jp), no-ri-ma-ki
**nori sembei** (Jp), no-ri sem-bei
**Normande, à la** (Fr), Nor-mande, à la
**Normande, sauce** (Fr), Nor-mande, sauce
**norrbottensost** (Sw), norr-bot-tens-ost
**no-salt herb blend** (US), no-salt herb blend
**nosh** (Jw), nosh
**noten** (Du), no-ten
**nötkött** (Sw), nöt-kött
**nøtter** (Nw), nøtt-er
**nougat** (US), nou-gat
**nouilles** (Fr), nou-illes
**nouvelle cuisine** (Fr), nou-velle cui-sine
**nozes** (Pg), no-zes
**nua phad prik** (Th), nua phad prik
**Nudeln** (Gr), Nu-deln
**Nudeln mit Kümmelkäse** (Gr), Nu-deln mit Küm-mel-kä-se

**Nudelsuppe mit Huhn** (Gr), Nu-del-sup-pe mit Huhn
**nudlar** (Sw), nud-lar
**nudle** (Cz), nu-dle
**nueces** (Sp), nu-e-ces
**nuez del Brasil** (Sp), nu-ez del Bra-sil
**nuoc cham** (Vt), nuoc cham
**nuoc mam** (Vt), nuoc mam
**Nurnbergerwurst** (Gr), Nurn-ber-ger-wurst
**Nussauflauf** (Gr), Nuss-auf-lauf
**Nüsse** (Gr), Nüs-se
**nut** (US), nut
**nuta** (Jp), nu-ta
**nutmeg** (US), nut-meg
**nutritive sweetener** (US), nu-tri-tive sweet-en-er
**nyama ya kaa** (Af-Swahili), n-ya-ma ya ka-a
**nyama ya kondoo** (Af-Swahili), n-ya-ma ya ko-ndo-o
**nyama ya kuchoma** (Af-Swahili), n-ya-ma ya ku-cho-ma
**nyama ya mbuzi** (Af-Swahili), n-ya-ma ya m-bu-zi
**nyama ya ndama** (Af-Swahili), n-ya-ma ya n-da-ma
**nyama ya ng'ombe** (Af-Swahili), n-ya-ma ya ng'o-mbe
**nyama ya nguruwe** (Af-Swahili), n-ya-ma ya n-gu-ru-we
**nybakt brød** (Nw), ny-bakt brød
**nyóu-pái** (Ch), nyóu-pái

**nyóu-ròu chīng tāng** (Ch), nyóu-ròu chīng tāng
**nyper** (Nw), ny-per
**nyponsoppa** (Sw), ny-pon-sop-pa

**nyre** (Da), ny-re
**nyrer** (Nw), ny-rer
**nysilt melk** (Nw), ny-silt melk
**nyúlpörkölt** (Hu), nyúl-pör-költ

# *O*

**oatmeal** (US), oat-meal
**oats** (US), oats
**obalované v housce** (Cz), o-ba-lo-va-né v ho-us-ce
**obed** (Rs), o-bed
**Oberskern** (Gr), O-bers-kern
**obiad** (Po), ob-iad
**obložené chlebíčky** (Cz), ob-lo-že-né chle-bíč-ky
**Obst** (Gr), Obst
**Obstkuchen** (Gr), Obst-ku-chen
**Obstsuppe** (Gr), Obst-sup-pe
**oca** (It), o-ca
**ocean perch** (US), o-cean perch
**ocet** (Cz, Po), o-cet
**ocha** (Jp), o-cha
**Ochsenbraten** (Gr), Och-sen-bra-ten
**Ochsenfleisch** (Gr), Och-sen-fleisch
**Ochsenlende** (Gr), Och-sen-len-de

**Ochsenmaulsalat** (Gr), Och-sen-maul-sa-lat
**Ochsenniere** (Gr), Och-sen-nie-re
**Ochsenschwanzsuppe** (Gr), Och-sen-schwanz-sup-pe
**Ochsenzunge** (Gr), Och-sen-zun-ge
**octopus** (US), oc-to-pus
**odamaki mushi** (Jp), o-da-ma-ki mu-shi
**oden** (Jp), o-den
**odoburu** (Jp), o-do-bu-ru
**oee kim chee** (Kr), o-ee kim chee
**oesters** (Du), oes-ters
**oeufs** (Fr), oeufs
**oeufs à la coque** (Fr), oeufs à la coque
**oeufs à la neige** (Fr), oeufs à la neige
**oeufs à la Richelieu** (Fr), oeufs à la Rich-e-lieu
**oeufs à la Russe** (Fr), oeufs à la Russe

*oeufs argenteuils* (Fr), oeufs ar-gen-teuils

*oeufs au plat* (Fr), oeufs au plat

*oeufs bénédictine* (Fr), oeufs bé-né-dic-tine

*oeufs Bercy* (Fr), oeufs Ber-cy

*oeufs brouillés* (Fr), oeufs brou-il-lés

*oeufs d'alose* (Fr), oeufs d'a-lo-se

*oeufs de caille au caviar* (Fr), oeufs de caille au ca-vi-ar

*oeufs de poisson* (Fr), oeufs de pois-son

*oeufs durs* (Fr), oeufs durs

*oeufs farcis* (Fr), oeufs far-cis

*oeufs frits* (Fr), oeufs frits

*oeufs mollets* (Fr), oeufs mol-lets

*oeufs pochés* (Fr), oeufs po-chés

*oeufs pochés en gelée* (Fr), oeufs po-chés en ge-lée

*oeufs Rossini* (Fr), oeufs Ros-si-ni

*of* (Jw-Israel), of

*offal* (GB), of-fal

*of sum-sum* (Jw-Israel), of sum-sum

*Offene Torte* (Gr), Of-fe-ne Tor-te

*ofu* (Jp), o-fu

*ogo* (Jp), o-go

*ogórek* (Po), o-gó-rek

*ogurtsy solionye v tykve* (Rs), o-gur-tsy so-li-o-nye v tyk-ve

*obagi* (Jp), o-ha-gi

*obrakyrsä* (Fi), oh-ra-kyr-sä

*obraryynipuuro* (Fi), oh-ra-ryy-ni-puu-ro

*obukaiset* (Fi), o-hu-ka-i-set

*oie* (Fr), oie

*oignon* (Fr), oi-gnon

*oignonade* (Fr), oi-gno-nade

*oil* (US), oil

*oiseau* (Fr), oi-seau

*oiseaux sans têtes* (Fr), oi-seaux sans têtes

*oj* (US), o-j

*oka* (Sp), o-ka

*Oka* (Ca), O-ka

*okame soba* (Jp), o-ka-me so-ba

*okaribayaki* (Jp), o-ka-ri-ba-ya-ki

*okashi* (Jp), o-ka-shi

*okayu* (Jp), o-ka-yu

*okonomiyaki* (Jp), o-ko-no-mi-ya-ki

*okoun* (Cz), o-ko-un

*okowa* (Jp), o-ko-wa

*okra* (US), ok-ra

*okroshka* (Rs), o-krosh-ka

*oksebaleragout* (Da), oks-e-ha-le-ra-gout

*oksekarbonade* (Nw), ok-se-kar-bo-na-de

*oksekjøtt* (Nw), ok-se-kjøtt

*oksekoedsuppe* (Da), oks-e-koed-sup-pe

*oksesteg* (Da), oks-e-steg

*oksestek* (Nw), ok-se-stek

*okurková omáčka* (Cz), o-kur-ko-vá o-máč-ka

*okurky* (Cz), o-kur-ky

*öl* (Sw), öl

*Öl* (Gr), Öl

*Öløl* (Nw), Öløl
*olaj* (Hu), o-laj
*olajbogyó* (Hu), o-laj-bo-gyó
*olebneena* (Rs), o-leh-nee-na
*olej* (Cz), o-lej
*óleo* (Pg), ó-le-o
*oleomargarine* (US), o-le-o-mar-ga-rine
*olie* (Du), o-lie
*oliebollen* (Du), o-lie-bol-len
*olijf* (Du), o-lij-f
*olio* (It), o-li-o
*olio santo* (It), o-li-o san-to
*oliva* (It, Pg, Sp), o-li-va
*olive* (US), ol-ive
*Oliven* (Gr), O-li-ven
*olivener* (Nw), o-li-ve-ner
*olive oil* (US), ol-ive oil
*oliver* (Sw), o-li-ver
*Olivet* (Fr), O-li-vet
*olivette* (It), o-li-vet-te
*olivi* (Fi), o-li-vi
*olivy* (Cz), o-li-vy
*oliwa* (Po), o-li-wa
*oliwki* (Po), o-liw-ki
*olja* (Sw), ol-ja
*olje* (Nw), ol-je
*öljy* (Fi), öl-jy
*olla podrida* (Sp), ol-la pod-ri-da
*øllebrød* (Da, Nw), øl-le-brød
*ölsardinen* (Gr), öl-sar-di-nen
*olut* (Fi), o-lut
*omáčka* (Cz), o-máč-ka
*omelet* (US), om-e-let
*omelete* (Pg), o-me-le-te
*omelett* (Nw), o-me-lett
*omelette à la confiture* (Fr), ome-lette à la con-fi-ture

*omelette au foie de volaille* (Fr), ome-lette au foie de vo-laille
*omelette au fromage* (Fr), ome-lette au fro-mage
*omelette au jambon* (Fr), ome-lette au jam-bon
*omelette au lard* (Fr), ome-lette au lard
*omelette aux fines herbes* (Fr), ome-lette aux fines herbes
*omelette aux girolles* (Fr), ome-lette aux girolles
*omelette basquaise* (Fr), ome-lette bas-quaise
*omelette bonne femme* (Fr), ome-lette bonne femme
*omelette nature* (Fr), ome-lette na-ture
*omelette norvégienne* (Fr), ome-lette nor-vé-gienne
*omelette parmentier* (Fr), ome-lette par-men-ti-er
*omelette provencale* (Fr), ome-lette pro-ven-cale
*omeletter* (Sw), o-me-let-ter
*omena* (Fi), o-me-na
*omenalumi* (Fi), o-me-na-lu-mi
*omenamunakas* (Fi), o-me-na-mu-na-kas
*omenapitko* (Fi), o-me-na-pit-ko
*omenosilli* (Fi), o-me-no-sil-li
*omlet* (Po), om-let
*omuretsu* (Jp), o-mu-ret-su
*onion* (US), on-ion

*onion flakes* (US), on-ion flakes
*onion rings* (US), on-ion rings
*ono* (Pl), o-no
*ontbijt* (Du), ont-bijt
*ooka* (Rs), oo-ka
*ooksoos* (Rs), ook-soos
*oolanin kakut* (Fi), oo-la-nin ka-kut
*oolong* (Ch), oo-long
*ooperavoileipä* (Fi), oo-pe-ra-vo-i-le-i-pä
*oostreetsi* (Rs), oo-stree-tsi
*ootka* (Rs), oot-ka
*opékané brambory* (Cz), o-pé-ka-né bram-bo-ry
*open-face* (US), o-pen-face
*operatårta* (Sw), o-pe-ra-tår-ta
*oplagt melk* (Nw), o-plagt melk
*orache* (US), or-ache
*orange* (US), or-ange
*orangeat* (Fr), or-an-geat
*orange drink* (US), or-ange drink
*orange juice* (US), or-ange juice
*orange mint* (US), or-ange mint
*orange oil* (US), or-ange oil
*orange pekoe* (US), or-ange pe-koe
*orange roughy* (US), or-ange rough-y
*ördek* (Tr), ör-dek
*oregano* (US), o-reg-a-no
*orekhovi pudding* (Rs), o-re-kho-vi pud-ding

*orenji jüsu* (Jp), o-ren-ji jü-su
*organen* (Du), or-ga-nen
*organic* (US), or-gan-ic
*orge* (Fr), orge
*orgeat* (Fr), or-ge-at
*origan* (Fr), o-ri-gan
*origano* (It), o-ri-ga-no
*ormer* (GB), or-mer
*orohova potica* (SC), o-ro-ho-va po-ti-ca
*ørred* (Da), ør-red
*ørret* (Nw), ør-ret
*ortaggio* (It), or-tag-gio
*ortolan* (Fr), or-to-lan
*orzechy* (Po), or-zec-hy
*orzo* (It), or-zo
*os à moelle* (Fr), os à moelle
*oseille* (Fr), o-seille
*osëtra* (Rs), o-sët-ra
*oshitashi* (Jp), o-shi-ta-shi
*osihtreena* (Rs), o-sih-tree-na
*osnovnoi biskvit* (Rs), os-nov-noi bis-kvit
*osnovnoi orekhovyi biskvit* (Rs), os-nov-noi o-re-khov-yi bis-kvit
*ossenhaas* (Du), os-sen-haas
*osso buco* (It), os-so bu-co
*ost* (Da, Nw. Sw), ost
*osterit* (Fi), os-te-rit
*østers* (Nw), øs-ters
*ostión* (Sp), os-ti-ón
*ostkaka* (Sw), ost-ka-ka
*ostra* (Sp), os-tra
*ostrá bořčice* (Cz), ost-rá hoř-či-ce
*ostra plana* (Pg), os-tra pla-na

*ostriche* (It), os-tri-che
*ostrige* (SC), o-stri-ge
*ostron* (Sw), o-stron
*ostryga* (Po), ost-ry-ga
*ostrý sýr* (Cz), ost-rý sýr
*ostsufflé* (Sw), ost-suf-flé
*öszibarack* (Hu), ö-szi-ba-rack
*osztriga* (Hu), oszt-ri-ga
*otvarnaia goviadina* (Rs), ot-var-na-ia go-via-di-na
*otvarnaia osetrina* (Rs), ot-var-na-ia o-se-tri-na
*otvarnoi kartofel* (Rs), ot-var-noi kar-to-fel
*ou* (Ch), ou
*oursins* (Fr), our-sins
*ouzo* (Gk), ou-zo
*ovčetina* (SC), ov-če-ti-na
*ovnstegt* (Da), ovn-stegt
*ovoce* (Cz), o-vo-ce
*ovocné knedliky* (Cz), o-voc-né kned-li-ky
*ovos* (Pg), o-vos
*ovos com fiambre* (Pg), o-vos com fi-am-bre
*ovos cozidos* (Pg), o-vos co-zi-dos
*ovos escalfados* (Pg), o-vos es-cal-fa-dos
*ovos estrelados* (Pg), o-vos es-tre-la-dos
*ovos fritos* (Pg), o-vos fri-tos

*ovos mexidos* (Pg), o-vos me-xi-dos
*ovos pochê* (Pg-Brazil), o-vos po-chê
*ovos quentes* (Pg), o-vos quen-tes
*owsianka* (Po), ow-sian-ka
*oxbringa* (Sw), ox-bring-a
*oxfilé* (Sw), ox-fi-lé
*Oxford sauce* (GB), Ox-ford sauce
*oxkött* (Sw), ox-kött
*oxrulader* (Sw), ox-ru-la-der
*oxstek* (Sw), ox-stek
*oxsvanssoppa* (Sw), ox-svans-sop-pa
*oxtail* (US), ox-tail
*oyako domburi* (Jp), o-ya-ko dom-bu-ri
*oyakodon* (Jp), o-ya-ko-don
*oyster* (US), oys-ter
*oyster cracker* (US), oys-ter crack-er
*oyster plant* (US), oys-ter plant
*oyster sauce* (US), oys-ter sauce
*oysters Bienville* (US), oys-ters Bi-en-ville
*oysters en brochette* (Fr), oys-ters en bro-chette
*oysters Rockefeller* (US), oys-ters Rock-e-fel-ler
*özgerinc* (Hu), öz-ge-rinc

# *P*

**paahtopaistivoileipä** (Fi), paah-to-pa-is-ti-voi-le-i-pä
**paalaeg** (Da), paa-laeg
**paan** (Ia), paan
**paani** (Ia), paa-ni
**paapar** (Ia), paa-par
**pääsiäisjuusto** (Fi), pää-si-ä-is-juus-to
**pääsiäisleipä** (Fi), pää-si-ä-is-le-i-pä
**pääsiäispasha** (Fi), pää-si-ä-is-pa-sha
**pabellón caraqueño** (Sp), pa-be-llón ca-ra-que-ño
**pacalpörkolt** (Hu), pa-cal-pör-kolt
**pacolt marhahus** (Hu), pa-colt mar-ha-hus
**paczki** (Po), pacz-ki
**paddlefish** (US), pad-dle-fish
**paella** (Sp), pa-el-la
**paellita** (Sp), pa-el-li-ta
**pærer** (Nw), pær-er
**paezinhos** (Pg-Brazil), pae-zi-nhos
**paglia e fieno** (It), pa-glia e fi-e-no
**pah jook** (Kr), pah jook
**pähkinäkakut** (Fi), päh-ki-nä-ka-kut

**pai** (Jp), pai
**pái dòu-fú ròu** (Ch), pái dòu-fú ròu
**pái-gŭ ròu** (Ch), pái-gŭ ròu
**paillard** (Fr), pai-llard
**paillassons** (Fr), pai-llas-sons
**paillettes au fromage** (Fr), pai-llettes au fro-mage
**pain** (Fr), pain
**pain à cacheter** (Fr), pain à cache-ter
**painappuru** (Jp), pa-in-ap-pu-ru
**pain aux noix** (Fr), pain aux noix
**pain bis** (Fr), pain bis
**pain complet** (Fr), pain com-plet
**pain de cuisine** (Fr), pain de cui-sine
**pain d'epice** (Fr), pain d'e-pice
**pain des algues** (Fr), pain des al-gues
**pain de seigle** (Fr), pain de seigle
**pain grillé** (Fr), pain gril-lé
**pain ordinaire** (Fr), pain or-di-naire
**pain perdu** (Fr), pain per-du

*paistettu kala* (Fi), pa-is-tet-tu ka-la

*paistettu metsälintu* (Fi), pa-is-tet-tu met-sä-lin-tu

*paistettu sianselka* (Fi), pa-is-tet-tu si-an-sel-ka

*paistettu silakka* (Fi), pa-is-tet-tu si-lak-ka

*paiusnaia* (Rs), pa-ius-na-ia

*paj* (Sw), paj

*pak choy* (Ch), pak choy

*paketti* (Fi), pa-ket-ti

*pakora* (Ia), pa-ko-ra

*palačinky* (Cz), pa-la-čin-ky

*palacsinta* (Hu), pa-la-csin-ta

*palacsintametélt* (Hu), pa-la-csin-ta-me-télt

*palak* (Ia), pa-lak

*palak bhurgi* (Ia), pa-lak bhur-gi

*palak murgh* (Ia), pa-lak murgh

*palak paneer* (Ia), pa-lak pa-neer

*palak raita* (Ia), pa-lak rai-ta

*palapaisti* (Fi), pa-la-pa-is-ti

*Palatschinken* (Gr), Pa-lat-schin-ken

*palée* (Fr), pa-lée

*paling in't groen* (Bl), pa-ling in't groen

*palline al cioccolato* (It), pal-li-ne al cioc-co-la-to

*palmier* (Fr), pal-mi-er

*palm oil* (US), palm oil

*palm sugar* (US), palm sug-ar

*palócleves* (Hu), pa-lóc-le-ves

*paloma* (Sp), pa-lo-ma

*palourdes* (Fr), pa-lourdes

*paltus s zelionym sousom* (Rs), pal-tus s ze-lio-nym sou-som

*pambacitos* (Sp), pam-ba-ci-tos

*Pampelmuse* (Gr), Pam-pel-mu-se

*Pampelmusensaft* (Gr), Pam-pel-mu-sen-saft

*pamplemousse* (Fr), pam-ple-mousse

*pamushki s chesnokom* (Rs), pa-mush-ki s ches-no-kom

*pan* (US), pan

*pan* (Jp, Sp), pan

*panaché* (Fr), pa-na-ché

*panaché au Roquefort* (Fr), pa-na-ché au Roque-fort

*panade* (Fr), pa-nade

*panado* (Pg), pa-na-do

*panais* (Fr), pa-nais

*panbroil* (US), pan-broil

*pancake* (US), pan-cake

*pancar* (Tr), pan-car

*pancetta* (It), pan-cet-ta

*pancit* (Ph), pan-cit

*pancit guisado* (Ph), pan-cit gui-sa-do

*pancit molo* (Ph), pan-cit mo-lo

*pan de bigos* (Sp), pan de bi-gos

*pan de centeno* (Sp), pan de cen-te-no

*pandekager* (Da), pan-de-ka-ger

*pan de maiz* (Sp), pan de ma-iz

*pan di spagna* (It), pan di spa-gna

*pandorato* (It), pan-do-ra-to
*pandowdy* (US), pan-dow-dy
*pan dulce* (Sp), pan dul-ce
*pane* (It), pa-ne
*pané* (Fr), pa-né
*pane a caponata* (It), pa-ne
a ca-po-na-ta
*pane caldo* (It), pa-ne cal-do
*panecillo* (Sp), pa-ne-ci-llo
*pane integrale* (It), pa-ne in-
te-gra-le
*panela* (Sp), pa-ne-la
*panelle* (It), pa-nel-le
*paner* (Fr), pa-ner
*panerat* (Sw), pa-ne-rat
*pane tostato e marmellata*
(It), pa-ne tos-ta-to e mar-
mel-la-ta
*panettone* (It), pan-et-to-ne
*panfish* (US), pan-fish
*panforte di Siena* (It), pan-
for-te di Sie-na
*panfry* (US), pan-fry
*pan giallo* (It), pan gial-lo
*pangsit goreng* (In), pang-sit
go-reng
*páng-syè* (Ch), páng-syè
*panier de crudites* (Fr), pan-
i-er de cru-di-tes
*paniert* (Gr), pa-niert
*panini* (It), pa-ni-ni
*panini di pasqua* (It), pa-ni-
ni di pas-qua
*panini imbotiti* (It), pa-ni-ni
im-bo-ti-ti
*panino gravido* (It), pa-ni-
no grav-i-do
*panir* (Ia), pa-nir
*panir tikka* (Ia), pa-nir tik-ka
*panko* (Jp), pan-ko

*panna* (It), pan-na
*panna cotta* (It), pan-na cot-
ta
*panna montata* (It), pan-na
mon-ta-ta
*pannato* (It), pan-na-to
*pannbiff med lök* (Sw),
pann-biff med lök
*pannekaka* (Nw), pan-ne-ka-
ka
*pannekoeken* (Du), pan-ne-
koe-ken
*pannequets* (Fr), pan-ne-
quets
*pannkakor med sylt* (Sw),
pann-ka-kor med sylt
*pannukakku* (Fi), pan-nu-
kak-ku
*panocha* (Sp), pa-no-cha
*panquecas* (Pg), pan-que-cas
*pan roasting* (US), pan roasting
*pansotti* (It), pan-sot-ti
*pantothenic acid* (US), pan-
to-then-ic ac-id
*pantua* (Ia), pan-tu-a
*panuchos* (Sp), pa-nu-chos
*panzanella* (It), pan-za-nel-la
*pão* (Pg), pão
*pão branco* (Pg), pão bran-co
*pão escuro* (Pg), pão es-cu-ro
*pao yü* (Ch), pao yü
*pãozinhos* (Pg), pã-o-zi-nhos
*pap* (Du), pap
*papa* (Sp), pa-pa
*papa* (Af-Swahili), pa-pa
*papa de cereal* (Pg), pa-pa
de ce-re-al
*papai* (Af-Swahili), pa-pa-i
*papain* (US), pa-pain

*papaja* (SC), pa-pa-ja

*papas chorriadas* (Sp), pa-pas cho-rri-a-das

*papas rellenas* (Mx), pa-pas re-lle-nas

*papatzul* (Sp), pa-pa-tzul

*papaw* (US), pa-paw

*papaya* (US), pa-pa-ya

*papillote, en* (Fr), pap-i-llote, en

*pappa al pomodoro* (It), pap-pa al po-mo-do-ro

*pappadam* (Ia), pap-pa-dam

*pappardelle* (It), pap-par-del-le

*pappardelle con lepre* (It), pap-par-del-le con le-pre

*pappardelle con porcini* (It), pap-par-del-le con por-ci-ni

*pappilan hätävara* (Fi), pap-pi-lan hä-tä-va-ra

*paprika* (US), pa-pri-ka

*paprika butter* (US), pa-pri-ka but-ter

*Paprika Hüner* (Gr-Austria), Pa-pri-ka Hü-ner

*paprikas* (Du), pa-pri-kas

*paprikás* (Hu), pa-pri-kás

*paprikás burgonya* (Hu), pa-pri-kás bur-go-nya

*paprikás csirke* (Hu), pa-pri-kás csir-ke

*paprikás mártás* (Hu), pa-pri-kás már-tás

*paprike* (SC), pa-pri-ke

*paprikový salát* (Cz), pap-ri-ko-vý sa-lát

*paquette* (Fr), pa-quette

*paradajz* (SC), pa-ra-da-jz

*Paradeiser* (Gr-Austria), Pa-ra-dei-ser

*Paradeissuppe* (Gr-Austria), Pa-ra-deis-sup-pe

*paradicsom* (Hu), pa-ra-di-csom

*paradicsommártás* (Hu), pa-ra-di-csom-már-tás

*Paraguay tea* (Sp), Par-a-guay tea

*Paranuss* (Gr), Pa-ra-nuss

*paratha* (Ia), pa-ra-tha

*parboil* (US), par-boil

*parch* (GB), parch

*parcha* (Sp), par-cha

*parchment paper* (US), par-ch-ment pa-per

*parcook* (US), par-cook

*pare* (US), pare

*pare-pare* (In), pa-re-pa-re

*pareve* (Jw), pa-re-ve

*parfait (Fr) par-fait*

*parilla, a la* (Sp), pa-ri-lla, a la

*pariloituvasikankaaryleet* (Fi), pa-ri-lo-i-tu-va-si-kan-kaa-ry-leet

*Paris-Brest* (Fr), Pa-ris-Brest

*Parisienne, à la* (Fr), Pa-ri-si-enne, à la

*Paris-Nice* (Fr), Pa-ris-Nice

*Parker House rolls* (US), Park-er House rolls

*parkins* (Sc), par-kins

*párky* (Cz), pár-ky

*parlies* (Sc), par-lies

*parma* (Cz), par-ma

*Parma ham* (It), Par-ma ham

*Parmentier* (Fr), Par-men-tier

*Parmesan* (US), Par-me-san

*parmigiana, alla* (It), par-mi-gia-na, al-la

*Parmigiano-Reggiano* (It), Par-mi-gia-no-Reg-gia-no

*päron* (Sw), pä-ron

*parposz* (Po), par-posz

*parsley* (US), pars-ley

*parsley root* (US), pars-ley root

*parsnip* (US), pars-nip

*partan* (Sc), par-tan

*partridge* (US), par-tridge

*parwal* (Ia), par-wal

*Pascal celery* (US), Pas-cal cel-er-y

*pasha* (Fi), pa-sha

*pashka* (Rs), pash-ka

*pashtet iz pechonki* (Rs), pash-tet iz pe-chon-ki

*pashtet iz ryby* (Rs), pash-tet iz ry-by

*pasilla* (US), pa-sil-la

*passas* (Pg), pas-sas

*passion fruit* (US), pas-sion fruit

*pasta* (Tr), pas-ta

*pasta* (It), pa-sta

*pasta al forno* (It), pa-sta al for-no

*pasta alla frutta* (It), pa-sta al-la frut-ta

*pasta all'uovo* (It), pa-sta al-l'uo-vo

*pasta asciutta* (It), pa-sta a-sciut-ta

*pasta con tonno* (It), pa-sta con ton-no

*pasta e fagioli* (It), pa-sta e fa-gio-li

*pasta filata* (It), pa-sta fi-la-ta

*pasta frolla* (It), pa-sta frol-la

*pasta in brodo* (It), pa-sta in bro-do

*pasta reale* (It), pa-sta re-a-le

*pasta sfoglia* (It), pa-sta sfo-glia

*paste* (It), pa-ste

*paste* (US), paste

*pastéis* (Pg), pas-té-is

*pastel* (Sp), pas-tel

*pastel de choclo* (Sp), pas-tel de cho-clo

*pastel de merluza* (Sp), pas-tel de mer-lu-za

*pasteles* (Sp), pas-tel-es

*pastelitos de boda* (Mx), pas-tel-i-tos de bo-da

*pastèque* (Fr), pas-tèque

*pasternak so smetanoi* (Rs), pas-ter-nak so sme-ta-noi

*pasteurization* (US), pas-teur-i-za-tion

*pasticciata* (It), pa-stic-cia-ta

*pasticcio di maccheroni* (It), pas-tic-cio di mac-che-ro-ni

*pastiera* (It), pa-stie-ra

*pastij* (Du), pas-tij

*pastille* (Fr), pas-ti-lle

*pastini* (It), pas-ti-ni

*pastirma* (Tr), pas-tir-ma

*pastitsio* (Gk), pa-stit-si-o

*pastoules* (Gk), pa-stou-les

*pastramá* (Bu), pas-tra-má

*pastrami* (US), pas-tra-mi

*pastrmka* (SC), pa-strm-ka

*pastry* (US), pas-try

*pastry cream* (US), pas-try cream

*pasty* (GB), pas-ty
*pasulj* (SC), pa-sulj
*pa sze ping kuo* (Ch), pa sze ping kuo
*paszteciki* (Po), pasz-te-ci-ki
*patakukko* (Fi), pa-ta-kuk-ko
*patata* (It, Sp), pa-ta-ta
*patata al forno* (It), pa-ta-ta al for-no
*patata bollita* (It), pa-ta-ta bol-li-ta
*patate* (Fr), pa-ta-te
*patate fritte* (It), fried potatoes.
*patates* (Gk), pa-ta-tes
*patates püresi* (Tr), pa-ta-tes pü-re-si
*pâté* (Fr), pâ-té
*pâte à pâté* (Fr), pâte à pâ-té
*pâté brisée* (Fr), pâ-té bri-sée
*pâté de campagne* (Fr), pâ-té de cam-pagn-e
*pâté de foie* (Fr), pâ-té de foie
*pâté de foie gras* (Fr), pâ-té de foie gras
*pâté en croûte* (Fr), pâ-té en croûte
*pâté en terrine* (Fr), pâ-té en ter-rine
*pâté maison* (Fr), pâ-té mai-son
*pâté molle* (Fr), pâ-té molle
*pâté pressée* (Fr), pâ-té pres-sée
*pâtes* (Fr), pâtes
*pâtisserie* (Fr), pâ-tiss-e-rie
*patina* (It), pa-ti-na
*patka* (SC), pat-ka
*patlican* (Tr), pat-li-can

*patlican kebabi* (Tr), pat-li-can ke-ba-bi
*patlicanli kebap* (Tr), pat-li-can-li ke-bap
*pato* (Pg, Sp), pa-to
*pat prik king* (Th), pat prik king
*patrijs* (Du), pa-trijs
*pattypan* (US), pat-ty-pan
*pau* (Ia), pau
*pauhi* (In), pau-hi
*pauhi tjha* (In), pau-hi tjha
*paupiettes* (Fr), pau-piettes
*paupiettes de sole* (Fr), pau-piettes de sole
*pavese* (It), pa-ve-se
*pavot* (Fr), pa-vot
*pawpaw* (US), paw-paw
*paximadakia* (Gk), pa-xi-ma-da-kia
*payasam* (Ia), pa-ya-sam
*payousnaya ikra* (Rs), pa-yous-na-ya i-kra
*payrot* (Jw-Israel), pay-rot
*paysanne, à la* (Fr), pay-sanne, à la
*pea* (US), pea
*peach* (US), peach
*peanut butter* (US), pea-nut but-ter
*peanut oil* (US), pea-nut oil
*pear* (US), pear
*pearl barley* (US), pearl bar-ley
*pearl sago* (US), pearl sa-go
*pearl tea* (Ch), pearl tea
*pease porridge* (GB), pease por-ridge
*pebernødder* (Da), pe-ber-nød-der

*pečen* (SC), pe-čen
*pečené* (Cz), pe-če-né
*pečené bovězí* (Cz), pe-če-né ho-vě-zí
*pečené na rožni* (Cz), pe-če-né na rož-ni
*pêche* (Fr), pêche
*pêches cardinal* (Fr), pêches car-di-nal
*pêches Melba* (Fr), pêches Mel-ba
*pechuga de pollo* (Sp), pe-chu-ga de po-llo
*pecivo* (SC), pe-ci-vo
*pecorino* (It), pec-o-ri-no
*pectin* (US), pec-tin
*pečurke* (SC), pe-čur-ke
*peel, to* (US), peel, to
*peel oil* (US), peel oil
*peertjes* (Du), peer-tjes
*peixe* (Pg), pe-i-xe
*peixe assado* (Pg), pe-i-xe as-sado
*peixe frito* (Pg), pe-i-xe fri-to
*pekmez* (SC), pek-mez
*Pelkartoffeln* (Gr), Pel-kar-tof-feln
*pemmican* (US), pem-mi-can
*penne* (It), pen-ne
*penne all' arrabbiata* (It), pen-ne al-l' ar-rab-bi-a-ta
*pennyroyal* (GB), pen-nyroy-al
*pepe* (It), pe-pe
*peper* (Du), pe-per
*peperonata* (It), pe-pe-ro-na-ta
*peperoni* (It), pe-pe-ro-ni
*pepino* (Pg), pe-pi-no
*pepino* (Sp), pe-pi-no

*pepitas* (Sp), pe-pi-tas
*peppar* (Sw), pep-par
*pepparkakor* (Sw), pep-par-ka-kor
*pepparrot* (Sw), pep-par-rot
*pepper* (Nw), pepp-er
*pepper, black* (US), pep-per, black
*peppercorns, green* (US), pep-per-corns, green
*pepperrot* (Nw), pepp-er-rot
*pepper, white* (US), pep-per, white
*pepř* (Cz), pepř
*pera* (Af-Swahili), pe-ra
*pera* (It), pe-ra
*pêra* (Pg), pê-ra
*perca* (Pg), per-ca
*percebe* (Sp), per-ce-be
*perche* (Fr), perche
*perdiz toledana* (Sp), per-diz tol-e-da-na
*perdreau* (Fr), per-dreau
*pere* (It), pe-re
*perigourdine* (Fr), pe-ri-gour-dine
*periwinkle* (GB), per-i-wink-le
*perkedel* (In), per-ke-del
*perna de carneiro* (Pg), per-na de car-ne-i-ro
*pernik* (Cz), per-ník
*pernuce* (It), per-nu-ce
*perogen* (Jw), pe-ro-gen
*persico* (It), per-si-co
*persika* (Sw), per-si-ka
*persil* (Fr), per-sil
*persilja* (Sw), per-sil-ja
*persillade* (Fr), per-sil-lade
*persille* (Nw), per-sill-e

*persillesovs* (Da), per-sil-le-sovs

*persimmon* (US), per-sim-mon

*peršun* (SC), per-šun

*perú* (Pg), pe-rú

*perzik* (Du), per-zik

*pesca* (It), pe-sca

*pescado guisado* (Sp), pes-ca-do gui-sa-do

*pesce* (It), pe-sce

*pesce spada* (It), pe-sce spa-da

*pesciolino* (It), pe-scio-li-no

*pêssego* (Pg), pê-sse-go

*pesto* (It), pes-to

*pesto Genovese* (It), pes-to Ge-no-ve-se

*pesto Romano* (It), pes-to Ro-ma-no

*petcha* (Jw), pet-cha

*Petersille* (Gr), Pe-ter-sil-le

*petit-beurre* (Fr), pe-tit-beurre

*petite marmite* (Fr), pe-tite mar-mite

*petit-lait* (Fr), pe-tit-lait

*petit mont blanc* (Fr), pe-tit mont blanc

*petit pain* (Fr), pe-tit pain

*petits pois* (Fr), pe-tits pois

*petrezselyem* (Hu), pe-tre-zse-lyem

*peya* (Af-Swahili), pe-ya

*peynir* (Tr), pey-nir

*peynirli pide* (Tr), pey-nir-li pi-de

*Pfannkuchen* (Gr), Pfann-ku-chen

*Pfeffer* (Gr), Pfef-fer

*Pfeffernüsse* (Gr), Pfef-fer-nüs-se

*Pfifferling* (Gr), Pfif-fer-ling

*Pfirsich* (Gr), Pfir-sich

*Pflaumen* (Gr), Pflaum-en

*Pflaumenmuss* (Gr), Pflaum-en-muss

*phalon ka pullao* (Ia), pha-lon ka pul-la-o

*pheasant* (US), pheas-ant

*pho* (Vt), pho

*phoa* (Ia), p-ho-a

*phosphates* (US), phos-phates

*phyllo* (Gk), phyl-lo

*picadinho* (Pg), pi-ca-di-nho

*picado* (Pg), pi-ca-do

*picante vinaigrette* (US), pi-can-te vin-ai-grette

*piccadillo* (Sp), pic-ca-di-llo

*piccalilli* (US), pic-ca-lil-li

*piccante* (It), pic-can-te

*piccioncini con risotto* (It), pic-ci-on-ci-ni con ri-sot-to

*piccione* (It), pic-ci-o-ne

*pichi* (Af-Swahili), pi-chi

*pickled mustard* (Ch), pick-led mus-tard

*pico de gallo* (Sp), pi-co de ga-llo

*pi-dàn* (Ch), pi-dàn

*pie* (US), pie

*pie à la mode* (US), pie à la mode

*pièce de résistance* (Fr), pièce de ré-sis-tance

*pieczeń wolwa* (Po), pie-czeń wol-wa

*pieds de porc* (Fr), pieds de porc

*piemontese, alla* (It), pie-mon-tese, al-la

*pie plant* (US), pie plant

*pieprz* (Po), piep-rz

*pigeon en cocotte* (Fr), pige-on en co-cotte

*pigeonneau* (Fr), pige-on-neau

*piggvar* (Nw, Sw), pigg-var

*pigments* (US), pig-ments

*pignoli* (It), pi-gno-li

*pignon* (Fr), pi-gnon

*piimä* (Fi), pii-mä

*piirakka* (Fi), pii-rak-ka

*pike* (US), pike

*pikkels* (Nw), pikk-els

*pilaf* (Tr), pi-laf

*pileća čorba* (SC), pi-le-ća čor-ba

*piletina* (SC), pi-le-ti-na

*piliç firinda* (Tr), pi-liç fi-rin-da

*pilipili manga* (Af-Swahili), pi-li-pi-li ma-nga

*pilipili shamba* (Af-Swahili), pi-li-pi-li sha-mba

*pilot biscuit* (US), pi-lot bis-cuit

*Pilze* (Gr), Pil-ze

*piman* (Jp), pi-man

*pimenta* (Pg), pi-men-ta

*pimenta do reino* (Pg), pi-men-ta do re-i-no

*pimenta verde* (Pg), pi-men-ta ver-de

*piments* (Fr), pi-ments

*pimiento* (Sp), pi-mi-en-to

*pimiento de Jamaica* (Sp), pi-mi-en-to de Ja-ma-i-ca

*pinaattiohukaiset* (Fi), pi-naat-ti-o-hu-ka-i-set

*pineapple* (US), pine-ap-ple

*pine nuts* (US), pine nuts

*ping-gwo* (Ch), ping-gwo

*ping-yú* (Ch), ping-yú

*pinnekjøtt* (Nw), pin-ne-kjøtt

*piñón* (Sp), pi-ñón

*pinole* (Mx), pi-no-le

*pinto bean* (US), pin-to bean

*piononos* (Sp), pi-o-no-nos

*pip* (US), pip

*pi-pá* (Ch), pi-pá

*pipérade* (Fr), pi-pé-rade

*pipián* (Sp), pi-pi-án

*pippuri* (Fi), pip-pu-ri

*pirinač* (SC), pi-ri-nač

*pirogi* (Rs), pi-ro-gi

*piroshki* (Rs), pi-rosh-ki

*pirzola* (Tr), pir-zo-la

*pisang goreng* (In), pi-sang go-reng

*piselli* (It), pi-sel-li

*pissenlits* (Fr), pis-sen-lits

*pista* (Ia), pis-ta

*pistaches* (Fr), pis-taches

*pistachio nuts* (US), pis-ta-chi-o nuts

*pisto manchego* (Sp), pis-to man-che-go

*pistou* (Fr), pis-tou

*pisztráng* (Hu), pisz-tráng

*pit* (US), pit

*pita* (SC), pi-ta

*pita sa orasima* (SC), pi-ta sa o-ra-si-ma

*pi-táu-kàn* (Ch), pi-táu-kàn

*pitepalt* (Sw), pi-te-palt

*pith* (US), pith

*pito-ja-joulupuuro* (Fi), pi-to-ja-jou-lu-puu-ro

*pivo* (Cz, SC), pi-vo

*pizza* (It), piz-za

*pizzaiola* (It), piz-zai-o-la

*plaice* (US), plaice

*pla kung* (Th), pla kung

*planking* (US), plank-ing

*plantain* (US), plan-tain

*plátano* (Sp), plá-ta-no

*plâteau de fromages* (Fr), plât-eau de fro-mag-es

*plättar* (Sw), plät-tar

*platýs* (Cz), pla-týs

*Plätzchen* (Gr), Plätz-chen

*pletionka s makom* (Rs), ple-ti-on-ka s ma-kom

*pleurote* (Fr), pleu-rote

*plísňový sýr* (Cz), plís-ňo-vý sýr

*pliushki* (Rs), pli-ush-ki

*pljeskavica* (SC), plje-ska-vi-ca

*plommer* (Nw), plomm-er

*plommon* (Sw), plom-mon

*plommonpudding* (Sw), plom-mon-pud-ding

*plommonspäckad fläsk-karré* (Sw), plom-mon-späc-kad fläsk-kar-ré

*plum* (US), plum

*plump, to* (US), plump, to

*plum sauce* (US), plum sauce

*poach* (US), poach

*poached eggs* (US), poached eggs

*poché* (Fr), po-ché

*pocherade ägg* (Sw), po-che-ra-de ägg

*pocheteau blanc* (Fr), po-che-teau blanc

*pod* (US), pod

*podded peas* (US), pod-ded peas

*podvarku* (SC), po-dvar-ku

*poêle, à la* (Fr), poêle, à la

*poffertjes* (Du), pof-fer-tjes

*pohovan* (SC), po-ho-van

*poi* (Pl), poi

*point, à* (Fr), point, à

*point d'asperges* (Fr), point d'as-per-ges

*poire* (Fr), poire

*poire à la Condé* (Fr), poire à la Con-dé

*poireaux* (Fr), poi-reaux

*poires belle Hélène* (Fr), poires belle Hé-lène

*pois* (Fr), pois

*pois à la francaise* (Fr), pois à la fran-caise

*pois cassés* (Fr), pois cas-sés

*pois chiche* (Fr), pois chiche

*pois et riz* (Fr), pois et riz

*poisson* (Fr), pois-son

*poisson d'eau douce* (Fr), pois-son d'eau douce

*poisson de mer* (Fr), pois-son de mer

*poitrine* (Fr), poi-trine

*poitrine de veau farcie* (Fr), poi-trine de veau far-cie

*poivrade* (Fr), poi-vrade

*poivre* (Fr), poivre

*poivre vert* (Fr), poivre vert

*poivron* (Fr), poi-vron

*polenta* (It), po-len-ta

*polenta e osei* (It), po-len-ta e o-sei

*polévky* (Cz), po-lév-ky
*Polish sausage* (US), Pol-ish sau-sage
*pollack* (US), pol-lack
*pollo* (It), pol-lo
*pollo* (Sp), po-llo
*pollo ai ferri* (It), pol-lo ai fer-ri
*pollo a la chilindrón* (Sp), po-llo a la chi-lin-drón
*pollo alla cacciatore* (It), pol-lo al-la ca-ccia-to-re
*pollo al vino bianco* (It), pol-lo al vi-no bian-co
*pollo arrosto* (It), pol-lo ar-ros-to
*pollo asado* (Sp), po-llo a-sa-do
*pollo fritto* (It), pol-lo frit-to
*Polonaise sauce* (Fr), Po-lo-naise sauce
*polpetta* (It), pol-pet-ta
*polpette* (It), pol-pet-te
*polpettone* (It), pol-pet-to-ne
*polpo* (It), pol-po
*pølse* (Nw), pøl-se
*polvo* (Pg), pol-vo
*polyunsaturated fats* (US), pol-y-un-sat-u-rat-ed fats
*polyunsaturated fatty acids* (US), pol-y-un-sat-u-rat-ed fat-ty ac-ids
*pomarańcza* (Po), po-ma-rań-cza
*pombe* (Af-Swahili), po-mbe
*pombo* (Pg), pom-bo
*pome fruits* (US), pome fruits
*pomegranate* (US), pome-gran-ate
*pomelo* (US), pom-e-lo

*pomeranč* (Cz), po-me-ranč
*pomidory* (Po), po-mi-do-ry
*pomme* (Fr), pomme
*pommes de terre* (Fr), pom-mes de terre
*pommes de terre à l'huile* (Fr), pommes de terre à l'huile
*pommes de terre duchesse* (Fr), pommes de terre du-chesse
*pommes frites* (Fr), pommes frites
*pommes gaufrettes* (Fr), pommes gau-frettes
*pommes purée* (Fr), pom-mes pu-rée
*pomodori* (It), po-mo-do-ri
*pomodori ripieni* (It), po-mo-do-ri ri-pie-ni
*pomodoro, al* (It), po-mo-do-ro, al
*pomorandža* (SC), po-mo-ran-dža
*pompano* (Sp), pom-pa-no
*pompano en papillote* (Fr), pom-pa-no en pa-pi-llote
*pompelmo* (It), pom-pel-mo
*pompelmoes* (Du), pom-pel-moes
*ponty* (Hu), ponty
*poori* (Ia), poo-ri
*popover* (US), pop-o-ver
*poppy seeds* (US), pop-py seeds
*porc* (Fr), porc
*porchetta* (It), por-chet-ta
*porcini* (It), por-ci-ni
*pôrco* (Pg), pôr-co
*porgy* (US), por-gy

**pork and beans** (US), pork and beans

**Porree** (Gr), Por-ree

**porri** (It), por-ri

**porridge** (GB), por-ridge

**portakal** (Tr), por-ta-kal

**porterhouse steak** (US), por-ter-house steak

**Porto** (Pg), Por-to

**Port Salut** (Fr), Port Sa-lut

**port wine** (US), port wine

**posillipo, alla** (It), po-sil-li-po, al-la

**posset** (GB), pos-set

**postej** (Da), po-stej

**potage** (Fr), po-tage

**potage clair** (Fr), po-tage clair

**potage crécy** (Fr), po-tage cré-cy

**potage crème** (Fr), po-tage crème

**potage crème de céleri** (Fr), po-tage crème de cé-le-ri

**potage crème d'epinards** (Fr), po-tage crème d'e-pi-nards

**potage de betterave** (Fr), po-tage de bet-te-rave

**potage Dubarry** (Fr), po-tage Du-bar-ry

**potage germiny** (Fr), po-tage ger-mi-ny

**potage parmentier** (Fr), po-tage par-men-tier

**potage St. Germain** (Fr), po-tage St. Ger-main

**potage tortue** (Fr), po-tage tor-tue

**potajes de garbanzos** (Sp), po-ta-jes de gar-ban-zos

**potassium** (US), po-tas-si-um

**potatis** (Sw), po-ta-tis

**potatismos** (Sw), po-ta-tis-mos

**potato** (US), po-ta-to

**potato bread** (US), po-ta-to bread

**potato buds** (US), po-ta-to buds

**potato chips** (US), po-ta-to chips

**potato flour** (US), po-ta-to flour

**pot-au-feu** (Fr), pot-au-feu

**poteter kokte** (Nw), po-te-ter kok-te

**poteter ovnstekte** (Nw), po-te-ter ovn-stek-te

**poteter stekte** (Nw), po-te-ter stek-te

**potetpuré** (Nw), po-tet-pu-ré

**potkas** (Sw), pot-kas

**pot liquor** (US), pot li-quor

**pot luck** (US), pot luck

**pot pie** (US), pot pie

**pot-posy** (Sc), pot-pos-y

**pot-roasting** (US), pot-roast-ing

**pots de crème au chocolate** (Fr), pots de crème au cho-co-late

**pottage** (GB), pot-tage

**pouding** (Fr), pou-ding

**poudre, en** (Fr), poudre, en

**poulard à la bongroise** (Fr), pou-lard à la hon-groise

**poulard à la vapeur** (Fr), pou-lard à la va-peur

**poule au pot** (Fr), poule au pot

**poulet** (Fr), pou-let

**poulet à la Marengo** (Fr), pou-let à la Ma-ren-go

**poulet chasseur** (Fr), pou-let chas-seur

**poulet de Bresse** (Fr), pou-let de Bresse

**poulet en cocotte** (Fr), pou-let en co-cotte

**poulet froid** (Fr), pou-let froid

**poulet rôti à l'estragon** (Fr), pou-let rô-ti à l'es-tra-gon

**poulette sauce** (Fr), pou-lette sauce

**poulpe** (Fr), poulpe

**poultry** (US), poul-try

**pound cake** (GB), pound cake

**poussin** (Fr), pous-sin

**powdered sugar** (US), pow-der-ed sug-ar

**pozole** (Sp), po-zo-le

**prairie oysters** (US), prai-rie oys-ters

**praline** (US), pra-line

**pranzo** (It), pran-zo

**pranzo di manzo** (It), pran-zo di man-zo

**prawn** (US), prawn

**preclík** (Cz), prec-lík

**preeo** (Th), pre-eo

**preheat** (US), pre-heat

**Preisselbeere** (Gr), Preis-sel-bee-re

**preliv za salutu** (SC), pre-liv za sa-lu-tu

**preserved eggs** (Ch), pre-served eggs

**preserves** (US), pre-serves

**pressed caviar** (US), pressed cav-i-ar

**Presskopf** (Gr), Press-kopf

**pressòkàvè** (Hu), pres-sò-kà-vè

**press peach** (US), press peach

**pressure cooker** (US), pres-sure cook-er

**presunto** (Pg), pre-sun-to

**pretzel** (US), pret-zel

**prezzemolo** (It), prez-ze-mo-lo

**prianiki tyl'skie** (Rs), pri-an-i-ki tyl'skie

**prickly pear** (US), prick-ly pear

**prik** (Th), prik

**prik haeng** (Th), prik haeng

**prik kee noo** (Th), prik kee noo

**prik mum** (Th), prik mum

**prima colazione** (It), pri-ma co-la-zi-o-ne

**primeiro almoĉo** (Pg), pri-me-i-ro al-mo-ĉo

**primeurs** (Fr), pri-meurs

**primo piatto** (It), pri-mo pi-at-to

**primost** (Nw), prim-ost

**printanière, à la** (Fr), prin-ta-nière, à la

**přírodní řízek** (Cz), pří-rod-ní ří-zek

**prixuelos** (Sp), pri-xu-e-los

**prodel** (Rs), pro-del

**profiterole** (Fr), pro-fit-er-ole

*profiteroles glacées au chocolate* (Fr), pro-fit-er-oles gla-cées au cho-co-late

*proja* (SC), pro-ja

*proof* (US), proof

*prosciutto* (It), pro-sciut-to

*prosciutto cotto* (It), pro-sciut-to cot-to

*prosná kaše* (Cz), pros-ná ka-še

*proteases* (US), pro-te-as-es

*protein* (US), pro-te-in

*proustille* (Fr), prous-tille

*provencale, à la* (Fr), pro-ven-cale, à la

*provola* (It), pro-vo-la

*Provolone* (It), Pro-vo-lo-ne

*pršuta* (SC), pr-šu-ta

*prugna* (It), pru-gna

*prugne cotte* (It), pru-gne cot-te

*prugne secche* (It), pru-gne sec-che

*pruim* (Du), pruim

*prune* (US), prune

*pruneau* (Fr), pru-neau

*pruneaux du pichet* (Fr), pru-neaux du pi-chet

*prune butter* (US), prune but-ter

*prune juice* (US), prune juice

*prünellen* (Gr), prü-nel-len

*prune whip* (US), prune whip

*pržen* (SC), pr-žen

*pržena jaja* (SC), pr-že-na ja-ja

*psaria ke thalasina* (Gk), psa-ria ke tha-la-si-na

*psarosupa* (Gk), psa-ro-su-pa

*pšeničný chléb* (Cz), pše-nič-ný chléb

*psito sto furno* (Gk), psi-to sto fur-no

*pstruh* (Cz), pstruh

*psyllium* (US), psyl-li-um

*ptcha* (Jw), pt-cha

*pua'a* (Pl), pu-a'a

*puchero* (Sp), pu-che-ro

*pudding* (US), pud-ding

*pudim* (Pg), pu-dim

*pudim flan* (Pg), pu-dim flan

*pudin de centollo* (Sp), pu-din de cen-to-llo

*pudink* (Cz), pu-dink

*pudin ya mayai* (Af-Swahili), pu-din ya ma-ya-i

*pu erh* (Ch), pu erh

*puerro* (Sp), pu-er-ro

*puff pastry* (US), puff past-ry

*puikot* (Fi), pu-i-kot

*puits d'amour* (Fr), puits d'a-mour

*pulgogi* (Kr), pul-go-gi

*pulla* (Fi), pul-la

*pullao* (Ia), pul-la-o

*pullao dilbahaar kajoo* (Ia), pul-la-o dil-ba-haar ka-joo

*pullet* (US), pul-let

*pulpeta* (Sp), pul-pe-ta

*pulpo* (Sp), pul-po

*pulse* (US), pulse

*pultost* (Nw), pult-ost

*pumelo* (US), pum-e-lo

*Pumpernickel* (Gr), Pum-per-nick-el

*pumpkin* (US), pump-kin

*pumpkin pie spice* (US), pump-kin pie spice

*punajuuret appelsiikastik-*

***keesa*** (Fi), pu-na-juu-ret ap-pel-sii-kas-tik-kee-sa

***punajuurikaalikeitto*** (Fi), pu-na-juu-ri-kaa-li-ke-it-to

***punajuuri salaatti*** (Fi), pu-na-juu-ri sa-laat-ti

***punaviinisilakka*** (Fi), pu-na-vii-ni-si-lak-ka

***punch*** (US), punch

***punjene paprike*** (SC), pu-nje-ne pa-pri-ke

***punsch*** (Sw), punsch

***puntas de filete*** (Sp), pun-tas de fil-e-te

***punto, al*** (It), pun-to, al

***puré de batatas*** (Pg), pu-ré de ba-ta-tas

***puré di patate*** (It), pu-ré di pa-ta-te

***purée*** (Fr), pu-rée

***purée de pois*** (Fr), pu-rée de pois

***purée de pommes de terre à***

***l'ail*** (Fr), pu-rée de pommes de terre à l'ail

***purée, to*** (US), pu-rée, to

***puri*** (Ia), pu-ri

***purjo*** (Sw), pur-jo

***purjokeitto*** (Fi), pur-jo-ke-it-to

***purpoo mulligatawny*** (Ia), pur-poo mul-li-ga-taw-ny

***puss pass*** (Nw), puss pass

***pú-táu*** (Ch), pú-táu

***Putenbraten*** (Gr), Pu-ten-bra-ten

***putera*** (SC), pu-te-ra

***putu*** (In), pu-tu

***puuroa*** (Fi), puu-ro-a

***puževi*** (SC), pu-že-vi

***pweza*** (Af-Swahili), pwe-za

***py mei fun*** (Th), py mei fun

***pyridoxine*** (US), py-ri-dox-ine

***pytt i panna*** (Sw), pytt i pan-na

***qabargah*** (Ia), qa-bar-gah

***qamar ad-din*** (Ar), qa-mar ad-din

***qamardin*** (Ar), qa-mar-din

***qar'*** (Ar), qar'

***qarnoun machi*** (Ar), qar-noun ma-chi

***qatayif*** (Ar), qa-ta-yif

***qishtah*** (Ar), qish-tah

***qorma*** (Ia), qor-ma

***qua*** (Vt), qua

*quaddid* (Ar), quad-did
*quadrucci* (It), qua-dru-cci
*quaglia* (It), quagl-ia
*quagliata* (It), quagl-ia-ta
*quaglie alla fiorentina* (It), quagl-ie al-la fio-ren-ti-na
*quaglie alla piemontese* (It), quagl-ie al-la pi-e-mon-te-se
*quaglie arrosto con polenta* (It), quagl-ie ar-ro-sto con po-len-ta
*quaqliette di vitello* (It), quaql-iet-te di vi-tel-lo
*quaglio* (It), quagl-io
*quahog* (US), qua-hog
*quail* (US), quail
*quandong* (Aa), quan-dong
*Quargel* (Gr), Quar-gel
*Quark* (Gr), Quark
*Quarkauflauf* (Gr), Quark-auf-lauf
*Quarkklösse* (Gr), Quark-klös-se
*Quarkkuchen* (Gr), Quark-ku-chen
*Quark mit Früchten* (Gr), Quark mit Früch-ten
*Quarkpfannkuchen* (Gr), Quark-pfann-ku-chen
*quarter* (US), quar-ter
*quartirolo* (It), quar-ti-ro-lo
*quarto de carneiro* (Pg), quar-to de car-ne-i-ro
*quasi* (Fr), quas-i
*quassia* (Du), quassi-a
*quatre épices* (Fr), quatre é-pices

*quatre mendiants* (Fr), quatre men-diants
*quatre-quarts* (Fr), quatre-quarts
*quattro stagioni* (It), quat-tro sta-gio-ni
*qubqub* (Ar), qub-qub
*queen conch* (US), queen conch
*queen of puddings* (GB), queen of pud-dings
*queen olive* (US), queen ol-ive
*queen scallop* (GB), queen scal-lop
*Queensland nut* (Aa), Queens-land nut
*queijadas* (Pg), que-i-ja-das
*queijinhos do ceu* (Pg), que-i-ji-nhos do ce-u
*queijo* (Pg), que-i-jo
*queijo de minas* (Pg), que-i-jo de mi-nas
*queijo de nata* (Pg), que-i-jo de na-ta
*queijo londrino* (Pg), que-i-jo lon-dri-no
*quelites* (Sp), que-li-tes
*quelites con chile ancho* (Mx), que-li-tes con chi-le an-cho
*quenelles* (Fr), que-nelles
*quenelles de brochet* (Fr), que-nelles de bro-chet
*quenelles de veau* (Fr), que-nelles de veau
*quente* (Pg), quen-te
*quesadilla* (Mx), que-sa-di-lla

**quesadillas de huitlacoche** (Mx), que-sa-di-llas de huit-la-co-che

**quesadillas de pollo** (Mx), que-sa-di-llas de po-llo

**queso** (Sp), que-so

**queso al horno** (Sp), que-so al hor-no

**queso añejo** (Mx), que-so añ-e-jo

**queso blanco** (Sp), que-so blan-co

**queso de almendra** (Mx), que-so de al-men-dra

**queso de cerdo** (Sp), que-so de cer-do

**queso del pais** (Sp), que-so del pa-is

**queso de prensa** (Sp), que-so de pren-sa

**queso enchilado** (Mx), que-so en-chi-la-do

**queso fresco** (Mx), que-so fres-co

**queso fundido** (Sp), que-so fun-di-do

**queso gallego** (Sp), que-so gal-le-go

**queso manchego** (Sp), que-so man-che-go

**queso rallado** (Sp), que-so ra-lla-do

**quetsche** (Fr), quetsche

**Quetschenkuchen** (Gr), Quet-schen-ku-chen

**queue** (Fr), queue

**queue de boeuf à l'auvergnate** (Fr), queue de boeuf à l'au-ver-gnate

**queue d'écrevisses gratinée** (Fr), queue d'é-cre-visses gra-ti-née

**queue de homard** (Fr), queue de ho-mard

**queue de porc** (Fr), queue de porc

**quiabo** (Pg), qui-a-bo

**quibebe** (Pg-Brazil), qui-be-be

**quiche** (Fr), quiche

**quiche Alsacienne** (Fr), quiche Al-sa-cienne

**quiche de jambon** (Fr), quiche de jam-bon

**quiche Lorraine** (Fr), quiche Lor-raine

**quick bread** (US), quick bread

**quick freeze** (US), quick freeze

**quignon** (Fr), qui-gnon

**quillet** (Fr), qui-llet

**quimbombo** (Cb), quim-bom-bo

**quin** (GB), quin

**quince** (US), quince

**quindims** (Pg), quin-dims

**quinine** (US), qui-nine

**quinnat** (US), quin-nat

**quinoa** (Sp), qui-no-a

**quinquina** (Fr), quin-qui-na

**quintal** (Fr), quin-tal

**quintoniles con chile mulato** (Mx), quin-to-ni-les con chi-le mu-la-to

**quisquillas** (Sp), quis-qui-llas

**Quitte** (Gr), Quit-te

**Quittenkonfekt** (Gr), Quit-ten-kon-fekt

**qursan** (Ar), qur-san

# R

*raakapihvi* (Fi), raa-ka-pih-vi
*raastelautanen* (Fi), raas-te-la-u-ta-nen
*raavilohi* (Fi), raa-vi-lo-hi
*rabacal* (Pg), ra-ba-cal
*rabakoz halkolbasz* (Hu), ra-ba-koz hal-kol-basz
*rabanetes* (Pg), ra-ba-ne-tes
*rábano* (Sp), rá-ba-no
*rábano picante* (Sp), rá-ba-no pi-can-te
*rabarbaro* (It), ra-bar-ba-ro
*rabarber* (Sw), ra-bar-ber
*rabarbersuppe* (Da), ra-bar-ber-sup-pe
*rabarbra* (Nw), ra-bar-bra
*rabbit* (US), rab-bit
*râble* (Fr), râble
*râble de lapereau* (Fr), râble de la-pe-reau
*rabo de toro* (Sp), ra-bo de to-ro
*rabo sueco* (Sp), ra-bo su-e-co
*racasse* (Fr), ra-casse
*raccoon* (US), rac-coon
*račići* (SC), ra-či-ći
*rack* (US), rack
*rack of lamb* (US), rack of lamb
*rack steaming* (US), rack steam-ing
*raclette* (Fr), rac-lette
*racuszki* (Po), ra-cusz-ki
*radicchio* (It), ra-dic-chio
*radicchio di Treviso* (It), ra-dic-chio di Tre-vi-so
*radicchio rosso di Verona* (It), ra-dic-chio ros-so di Ve-ro-na
*Radieschen* (Gr), Ra-dies-chen
*radiki* (Gk), ra-di-ki
*radikia me ladi* (Gk), ra-di-kia me la-di
*radis* (Fr), ra-dis
*radis beurre* (Fr), ra-dis beu-rre
*radish* (US), rad-ish
*rädisor* (Sw), rä-di-sor
*rafano* (It), ra-fa-no
*rågbröd* (Sw), råg-bröd
*raggmunker* (Sw), ragg-mun-ker
*ragi* (Ia), ra-gi
*ragi dosas* (Ia), ra-gi dos-as
*ragnons de veau* (Fr), ra-gnons de veau
*ragoût* (Fr), ra-goût

**ragoût chipolata** (Fr), ra-goût chi-po-la-ta

**ragout des pattes et bou-lettes** (Fr), ra-gout des pat-tes et bou-lettes

**ragù** (It), ra-gù

**ragù di fegatini** (It), ra-gù di fe-ga-ti-ni

**rahkapiirakat** (Fi), rah-ka-pii-ra-kat

**rahkapiirakka** (Fi), rah-ka-pii-rak-ka

**rahkatorttu** (Fi), rah-ka-tort-tu

**Rahm** (Gr), Rahm

**raie** (Fr), raie

**raie au beurre noir** (Fr), raie au beurre noir

**raie fritte** (Fr), raie fritte

**raifort** (Fr), rai-fort

**raifort sauce** (Fr), rai-fort sauce

**rainbow trout** (US), rain-bow trout

**raisin bark** (US), rai-sin bark

**raisin sec** (Fr), rai-sin sec

**raisins** (US), rai-sins

**raisins** (Fr), rai-sins

**raisu** (Jp), rai-su

**raita** (Ia), ra-i-ta

**raiton** (Fr), rai-ton

**raja gladka** (Po), ra-ja glad-ka

**rajas con crema** (Sp), ra-jas con crem-a

**rajas dalna** (Ia), raj-as dal-na

**rajas de chili poblano** (Sp), ra-jas de chi-li po-bla-no

**rajma dal** (Ia), raj-ma dal

**rajská omáčka** (Cz), ra-j-ská o-máč-ka

**rak** (SC), rak

**räkor** (Sw), rä-kor

**rakørret** (Nw), ra-kør-ret

**räksallad** (Sw), räk-sal-lad

**rakott burgonya** (Hu), ra-kott bur-gon-ya

**rakott káposzta** (Hu), ra-kott ká-posz-ta

**rakott metélt** (Hu), ra-kott me-télt

**ral-la** (Ar), ral-la

**rambutan** (In), ram-bu-tan

**ramekin** (US), ram-e-kin

**ramen** (Jp), ra-men

**ramequin** (Fr), rame-quin

**ramps** (US), ramps

**ranchero sauce** (US), ranch-er-o sauce

**rancidity** (US), ran-cid-i-ty

**rane** (It), ra-ne

**rane in guazetto** (It), ra-ne in gua-zet-to

**rántotta** (Hu), rán-tot-ta

**rántotta zöldpaprikával** (Hu), rán-tot-ta zöld-pap-ri-ká-val

**rántott ponty** (Hu), rán-tott ponty

**rántott sértes borda** (Hu), rán-tott sér-tes bor-da

**rapa** (It), ra-pa

**rapanelli** (It), ra-pa-nel-li

**rapaperikiisseli** (Fi), ra-pa-per-i-kiis-se-li

**rapa svedese** (It), ra-pa sve-de-se

**rape** (US), rape

**rape** (GB), rape

*rapphöna* (Sw), rapp-hö-na
*rapphøns* (Nw), rapp-høns
*rapukeitto* (Fi), ra-pu-ke-it-to
*rårakor* (Sw), rå-ra-kor
*rare* (US), rare
*rarebit* (US), rare-bit
*rasam* (Ia), ra-sam
*rascasse rouge* (Fr), ras-casse rouge
*rasgulas* (Ia), ras-gul-as
*rasher* (GB), rash-er
*rasomalai* (Ia), ra-so-ma-lai
*raspberry* (US), rasp-ber-ry
*rasstegai* (Rs), ras-ste-gai
*rasstegai s ryboi* (Rs), ras-ste-gai s ry-boi
*rassypchataia grechnevaia kasha* (Rs), ras-syp-cha-ta-ia grech-ne-va-ia ka-sha
*rassypchatoe testo* (Rs), ras-syp-cha-toe tes-to
*ratafia* (GB), ra-ta-fi-a
*ratafias* (It), ra-ta-fi-as
*ratatouille* (Fr), ra-ta-tou-ille
*Rauhreif* (Gr), Rauh-reif
*rauwe biefstuk* (Du), rau-we bief-stuk
*rauwe haring* (Du), rau-we ha-ring
*rauwkostsla* (Du), rauw-kost-sla
*rava idli* (Ia), ra-va id-li
*ravani* (Gk), ra-va-ni
*ravenelli* (It), ra-ve-nel-li
*ravigote* (Fr), ra-vi-gote
*ravioli* (It), ra-vi-o-li
*ravut* (Fi), ra-vut
*rawon* (In), ra-won
*ray* (US), ray
*rayta* (Ia), ra-y-ta

*razmaznia* (Rs), raz-maz-nia
*ražnjići* (SC), raž-nji-ći
*razor clam* (GB), ra-zor clam
*rebā* (Jp), re-bā
*rebarbara es eper lekvar* (Hu), re-bar-ba-ra es e-per lek-var
*Rebhuhn* (Gr), Reb-huhn
*Rebhuhner mit Weintrauben* (Gr), Reb-huh-ner mit Wein-trau-ben
*Reblochon* (Fr), Re-blo-chon
*reçel* (Tr), re-çel
*recheio* (Pg), re-che-i-o
*rečni rak* (SC), reč-ni rak
*reconstitute* (US), re-con-sti-tute
*red bean* (US), red bean
*red bean sauce* (US), red bean sauce
*red beans and rice* (US), red beans and rice
*red cabbage* (US), red cab-bage
*red cooked* (US), red cooked
*red currant* (GB), red-cur-rant
*reddik* (Nw), red-dik
*red-eye gravy* (US), red-eye gra-vy
*redfish* (US), red-fish
*red flannel hash* (US), red flan-nel hash
*red herring* (GB), red her-ring
*ředkvička* (Cz), řed-kvič-ka
*red miso* (Jp), red mi-so
*red snapper* (US), red snap-per
*reduce* (US), re-duce

**reduced-calorie** (US), re-duced-cal-o-rie

**reduced-fat** (US), re-duced-fat

**reduced-sodium** (US), re-duced-so-di-um

**reduction** (US), re-duc-tion

**red wines** (US), red wines

**refined** (US), re-fined

**refined cereal** (US), re-fined ce-re-al

**refined sugar** (US), re-fined sug-ar

**reerug** (Du), ree-rug

**refogado** (Pg), re-fo-ga-do

**refresh** (US), re-fresh

**refried beans** (US), re-fried beans

**refritos** (Mx), re-fri-tos

**réglisse** (Fr), ré-glisse

**Rehbraten** (Gr), Reh-bra-ten

**Rebrücken** (Gr), Reh-rü-cken

**Rebrücken mit Rotwein Sose** (Gr), Reh-rü-cken mit Rot-wein So-se

**rehydrate** (US), re-hy-drate

**Reibekuchen** (Gr), Rei-be-ku-chen

**Reiberdatschi** (Gr), Rei-ber-dat-schi

**reikaleipa** (Fi), re-i-ka-le-i-pa

**reine** (Fr), reine

**reine-claude** (Fr), reine-claude

**reine de saba** (Fr), reine de sa-ba

**Reis** (Gr), Reis

**rejer** (Da), re-jer

**reker** (Nw), re-ker

**relevé** (Fr), re-le-vé

**relevée, sauce** (Fr), re-le-vée, sauce

**religieuse** (Fr), re-li-gieuse

**relish** (US), rel-ish

**rellenong alimango** (Ph), rel-le-nong a-li-man-go

**remis** (In), re-mis

**remis besar** (In), re-mis be-sar

**rémol** (Sp), ré-mol

**Remoladensosse** (Gr), Re-mo-la-den-sos-se

**remoulade** (Fr), re-mou-lade

**rempah** (In), rem-pah

**rempah-rempah udang** (In), rem-pah-rem-pah u-dang

**rempejek** (In), rem-pe-jek

**rempejek bayam** (In), rem-pe-jek ba-yam

**rempejek kedele** (In), rem-pe-jek ke-de-le

**renaissance, à la** (Fr), re-nai-ssance, à la

**render** (US), rend-er

**renkon** (Jp), ren-kon

**rennet** (US), ren-net

**rennet pudding** (GB), ren-net pud-ding

**rennin** (US), ren-nin

**rensdyrstek** (Nw), rens-dyr-stek

**renstek** (Sw), ren-stek

**repa s sousom iz malagi** (Rs), re-pa s sou-som iz ma-la-gi

**repôlho** (Pg), re-pô-lho

**rétesh** (Hu), ré-tesh

**Rettich** (Gr), Re-ttich

**retasu** (Jp), re-ta-su

*reteges sonkatorta* (Hu), re-te-ges son-ka-tor-ta
*retinol* (US), ret-i-nol
*retsina* (Gk), ret-si-na
*Reuben sandwich* (US), Reu-ben sand-wich
*revani* (Gk), re-va-ni
*revbensspjäll* (Sw), rev-bens-spjäll
*reveň* (Cz), re-veň
*reverdine* (Fr), re-ver-dine
*revithosupa* (Gk), re-vi-tho-su-pa
*rezanci* (SC), re-zan-ci
*Rhine wines* (Gr), Rhine wines
*Rhône wines* (Fr), Rhône wines
*rhubarb* (US), rhu-barb
*riabchiki pechonye v gline, na okhote* (Rs), riab-chi-ki pe-cho-nye v gli-ne, na o-kho-te
*rib chops* (US), rib chops
*rib eye* (US), rib eye
*riboflavin* (US), ri-bo-fla-vin
*ribollita* (It), ri-bol-li-ta
*rib roast* (US), rib roast
*ricciarelli* (It), ric-cia-rel-li
*riccio* (US), ric-ci-o
*rice* (US), rice
*rice miso* (Jp), rice mi-so
*ricer* (US), ric-er
*rice vinegar* (US), rice vin-e-gar
*rice wines* (Jp), rice wines
*ricotta* (It), ri-cot-ta
*ricotta salata* (It), ri-cot-ta sa-la-ta
*rieska* (Fi), ries-ka

*Riesling* (Gr), Ries-ling
*rigaglie* (It), ri-ga-glie
*rigani* (Gk), ri-ga-ni
*rigatoni* (It), rig-a-to-ni
*rigatoni al forno col ragù* (It), ri-ga-to-ni al for-no col ra-gù
*rigatoni con salsiccia* (It), ri-ga-to-ni con sal-sic-cia
*Rigó Jancsi* (Hu), Ri-gó Jan-csi
*riisi* (Fi), rii-si
*riisimakkara* (Fi), rii-si-mak-ka-ra
*riisipuuro* (Fi), rii-si-puu-ro
*riivinkropsu* (Fi), rii-vin-krop-su
*rijst* (Du), rijst
*rijstpap* (Du), rijst-pap
*rijsttafel* (Du), rijst-ta-fel
*rillauds* (Fr), ri-llauds
*rillettes* (Fr), ri-llettes
*rilletes d'oie* (Fr), ri-lletes d'oie
*rilletes de porc* (Fr), ri-lletes de porc
*rillons* (Fr), ri-llons
*rimmad skinka* (Sw), rim-mad skin-ka
*Rinderrouladen* (Gr), Rin-der-rou-la-den
*Rindfleisch* (Gr), Rind-fleisch
*Rindfleischkochwurst* (Gr), Rind-fleisch-koch-wurst
*Ringelblume* (Gr), Rin-gel-blume
*ringo* (Jp), rin-go
*riñones* (Sp), ri-ño-nes
*riñones al Jerez* (Sp), ri-ño-nes al Je-rez

*riñones de ternera* (Sp), ri-
ño-nes de ter-ne-ra
*rins* (Pg), rins
*rins de vitela* (Pg), rins de vi-
te-la
*Rioja* (Sp), Ri-o-ja
*ripa* (Sw), ri-pa
*ripieni* (It), ri-pi-eni
*ripieni di castagna* (It), ri-
pi-eni di cas-ta-gna
*ripieno* (It), ri-pi-eno
*rips* (Nw), rips
*ris* (Da, Nw, Sw), ris
*ris* (Rs), ris
*ris de veau* (Fr), ris de veau
*ris de veau archiduc* (Fr), ris
de veau arch-i-duc
*risgrynsgröt* (Sw), ris-gryns-
gröt
*risi e bisi* (It), ri-si e bi-si
*riso* (It), ri-so
*riso al limone* (It), ri-so al li-
mo-ne
*risotto* (It), ri-sot-to
*risotto a Barolo* (It), ri-sot-to
a Ba-ro-lo
*risotto alla milanese* (It), ri-
sot-to al-la mi-la-ne-se
*risotto al limone* (It), ri-sot-
to al li-mo-ne
*risotto con le seppie* (It), ri-
sot-to con le sep-pie
*risotto con scampi* (It), ri-
sot-to con scam-pi
*risotto e funghi* (It), ri-sot-to
e fun-ghi
*risotto marinaio* (It), ri-sot-
to ma-ri-naio
*risotto tortino* (It), ri-sot-to
tor-ti-no

*risovye kotlety* (Rs), ris-ovye
kot-le-ty
*rissóis* (Pg), ris-só-is
*rissole* (Fr), ris-sole
*rissoler* (Fr), ris-sol-er
*ristet brød* (Nw), rist-et
brød
*ristras* (Sp), ris-tras
*riz* (Fr), riz
*riz à l'impératrice* (Fr), riz à
l'im-pé-ra-trice
*riz étuvé au beurre* (Fr), riz
é-tu-vé au beurre
*riža* (SC), ri-ža
*rizi* (Gk), ri-zi
*rizogalo* (Gk), ri-zo-ga-lo
*rizs* (Hu), rizs
*rizsfelfújt* (Hu), rizs-fel-fújt
*roast* (Fr), roast
*roaster* (US), roast-er
*robalo* (Pg), ro-ba-lo
*robata* (Jp), ro-ba-ta
*robatayaki* (Jp), ro-ba-ta-ya-
ki
*Robert sauce* (Fr), Ro-bert
sauce
*robiola* (It), ro-bi-o-la
*Rock Cornish hen* (GB),
Rock Corn-ish hen
*rocket cress* (US), rock-et
cress
*rockfish* (US), rock-fish
*rockling* (Au), rock-ling
*Rock Point* (US), Rock Point
*rocky road* (US), rock-y road
*rodaballo* (Sp), ro-da-ba-llo
*rødbeter* (Nw), rød-be-ter
*rodbedesalat* (Da), rod-
bede-sa-lat
*rödbeter* (Nw), röd-be-ter

**rödbetor** (Sw), röd-be-tor
**rodekool** (Du), ro-de-kool
**rode wijn** (Du), ro-de wijn
**rødgrød med fløde** (Da),
rød-grød med flø-de
**rødgrøt** (Nw), rød-grøt
**rodovalho** (Pg), ro-do-va-lho
**rödspätta** (Sw), röd-spät-ta
**rødspette** (Nw), rød-spet-te
**rödvin** (Sw), röd-vin
**rodzynki** (Po), ro-dzyn-ki
**roe** (GB), roe
**roereieren** (Du), roer-ei-e-
ren
**rogan josh** (Ia), ro-gan josh
**roggebrood** (Du), rog-ge-
brood
**rognon** (Fr), ro-gnon
**rognoncini trifolati** (It), ro-
gnon-ci-ni tri-fo-la-ti
**rognoni** (It), ro-gno-ni
**rognoni di vitello** (It), ro-
gno-ni di vi-tel-lo
**rognons de veau** (Fr), ro-
gnons de veau
**rognons de veau à la mout-
arde** (Fr), ro-gnons de
veau à la mou-tarde
**roblík** (Cz), roh-lík
**robliky** (Jw), roh-li-ky
**Rohwurst** (Gr), Roh-wurst
**rojoes comino** (Pg), ro-jo-es
co-mi-no
**røket laks** (Nw), rø-ket laks
**rollatini** (It), rol-la-ti-ni
**rollatini di vitella al pomo-
doro** (It), rol-la-ti-ni di vi-
tel-la al po-mo-do-ro
**roll, to** (US), roll, to

**rolled Boston butt** (US),
rolled Bos-ton butt
**rolled double sirloin roast**
(US), rolled dou-ble sir-
loin roast
**rolled lamb shoulder** (US),
rolled lamb shoul-der
**rolled oats** (US), rolled oats
**rolled rib** (US), rolled rib
**rolled rump** (US), rolled
rump
**rollmops** (Gr), roll-mops
**roll-ups** (US), roll-ups
**rolpens** (Du), rol-pens
**roly-poly pudding** (GB), ro-
ly-po-ly pud-ding
**rom** (Sw), rom
**Romadur** (Gr), Ro-ma-dur
**romagnola, alla** (It), ro-ma-
gno-la, al-la
**romaine** (US), ro-maine
**romana, alla** (It), ro-ma-na,
al-la
**Romano** (It), Ro-man-o
**Romanello** (It), Ro-ma-nel-lo
**romarin** (Fr), ro-ma-rin
**rombo chiodato** (It), rom-bo
chi-o-da-to
**rombo liscio** (It), rom-bo lis-
ci-o
**romeritos** (Sp), ro-me-ri-tos
**romescu** (Sp), ro-mes-cu
**rømmergrøt** (Nw), røm-mer-
grøt
**rompope** (Sp), rom-po-pe
**rookvlees** (Du), rook-vlees
**room** (Du), room
**roomsoezen** (Du), room-
soe-zen
**rooz** (Ar), rooz

*ropa vieja* (Mx), ro-pa vi-e-ja
*Roquefort* (Fr), Roque-fort
*Roquefort gougères* (Fr), Roque-fort gou-gères
*røræg* (Da), rør-æg
*rosbif* (Fr, It, Sp), ros-bif
*rosbife* (Pg), ros-bi-fe
*rosca* (Pg), ros-ca
*rosca de los Reyes* (Mx), ros-ca de los Rey-es
*rosé* (US), ro-sé
*rose hips* (US), rose hips
*rosemary* (GB), rose-mar-y
*rosenkål* (Nw), ro-sen-kål
*Rosenkohl* (Gr), Ro-sen-kohl
*rose petals* (US), rose pet-als
*rosette* (It), ro-set-te
*rosette de Lyons* (Fr), ro-sette de Ly-ons
*rose water* (US), rose wa-ter
*rosé wine* (Fr), ro-sé wine
*rosin* (Nw), ro-sin
*rosmarino* (It), ros-ma-ri-no
*rosól* (Po), ro-sól
*rosolli* (Fi), ro-sol-li
*rosollikastike* (Fi), ro-sol-li-kas-ti-ke
*Rossini* (Fr), Ros-si-ni
*rosso* (It), ros-so
*rostat bröd* (Sw), ro-stat bröd
*rostbiff* (Nw, Sw), rost-biff
*Rostbraten* (Gr), Rost-bra-ten
*rösten* (Gr), rös-ten
*rostelyos* (Hu), ros-te-lyos
*Rostkartoffeln* (Gr), Rost-kar-tof-feln
*rösti* (Gr-Swiss), rös-ti
*rosto* (Gk, Tr), ros-to
*roston* (It), ros-ton

*rōsuto bīfu* (Jp), rō-su-to bī-fu
*Röte Grutze* (Gr), Rö-te Gru-tze
*roti* (In), ro-ti
*roti* (Ia), ro-ti
*rôti* (Fr), rô-ti
*roti bolu empuh* (In), ro-ti bo-lu em-puh
*roti bolu keras* (In), ro-ti bo-lu keras
*rôti de porc au lait* (Fr), rô-ti de porc au lait
*rôti de veau vinaigrette* (Fr), rô-ti de veau vi-nai-grette
*roties* (Fr), ro-ties
*roti hitam* (In), ro-ti hi-tam
*roti putih* (In), ro-ti pu-tih
*rotisserie* (Fr), ro-tis-se-rie
*Rötkohl* (Gr), Röt-kohl
*Rötkohl mit Apfeln* (Gr), Röt-kohl mit Ap-feln
*rotmos* (Sw), rot-mos
*rotoli di manzo* (It), ro-to-li di man-zo
*rotoli di vitello* (It), ro-to-li di vi-tel-lo
*rotselleri* (Sw), rot-sel-le-ri
*Rotwein* (Gr), Rot-wein
*ròu* (Ch), ròu
*ròu-bāu* (Ch), ròu-bāu
*rouelle de citron* (Fr), rouelle de ci-tron
*rouelle de veau* (Fr), rouelle de veau
*rouget* (Fr), rou-get
*rouget à la Bordelaise* (Fr), rou-get à la Bor-de-laise
*rouget de roche* (Fr), rou-get de roche

**rouille** (Fr), rou-ille
**roulade** (Fr), rou-lade
**Rouladen** (Gr), Rou-la-den
**round of beef** (US), round of beef
**ròu-pyàn** (Ch), ròu-pyàn
**ròu wán** (Ch), ròu wán
**roux** (US), roux
**rovita jaja** (SC), ro-vi-ta ja-ja
**rowanberry** (GB), row-an-ber-ry
**royal** (Gk), roy-al
**Royal Brabant** (Bl), Roy-al Bra-bant
**royale, á la** (Fr), roy-ale, á la
**rozbif** (Tr), roz-bif
**rozbratie z cebula** (Po), roz-bra-tie z ce-bu-la
**roze** (Ar), roze
**Rube** (Gr), Ru-be
**rubiyan** (Ar), ru-bi-yan
**rubra** (It), ru-bra
**ruby port** (GB), ru-by port
**rue** (GB), rue
**Rüebli** (Gr-Switzerland), Rüe-bli
**rugbrød** (Nw), rug-brød
**rǔ-gē** (Ch), rǔ-gē
**Rübreier** (Gr), Rühr-ei-er
**ruiskorppu** (Fi), ru-is-korp-pu
**rujak** (In), ru-jak
**rum** (GB), rum
**rumbledethumps** (Sc), rum-ble-de-thumps
**rump roast** (US), rump roast
**runderrollade** (Du), run-der-rol-la-de
**rundstykke** (Nw), rund-styk-ke

**rundvlees** (Du), rund-vlees
**runner bean** (US), run-ner bean
**rusalda** (Sp), ru-sal-da
**rusk** (US), rusk
**ruskistettua** (Fi), rus-kis-tet-tu-a
**russel** (Jw), rus-sel
**russet** (US), rus-set
**Russian dressing** (US), Rus-sian dress-ing
**russin** (Sw), ru-ssin
**rusty dab** (US), rust-y dab
**rutabaga** (US), ru-ta-ba-ga
**rūz** (Ar), rūz
**rūz ahbyahd** (Ar), rūz ahb-yahd
**rūz billabun** (Ar), rūz bil-la-bun
**ryba** (Rs), ry-ba
**rybí filé** (Cz), ry-bí fi-lé
**rybiz** (Cz), ry-biz
**rybnaya ikra** (Rs), ryb-na-ya i-kra
**ryby** (Po), ry-by
**ryby duzone w smietani** (Po), ry-by du-zo-ne w smie-ta-ni
**rye flour** (US), rye flour
**rye whiskey** (US), rye whis-key
**rype** (Da,Nw), ry-pe
**ryyppy** (Fi), ry-yp-py
**ryytikalasalaatti** (Fi), ry-y-ti-ka-la-sa-laat-ti
**ryż** (Po), ryż
**ryz do legumin** (Po), ryz do le-gu-min
**rýže** (Cz), rý-že
**rzodkiewka** (Po), rzod-kiew-ka

# S

**saag** (Ia), saag
**Saanen** (Gr), Saa-nen
**saba** (Jp), sa-ba
**sábalo** (Sp), sá-ba-lo
**sabanegh** (Ar), sa-ba-negh
**saba no misoni** (Jp), sa-ba no mi-so-ni
**saba no suzuke** (Jp), sa-ba no su-zu-ke
**sabayon** (Fr), sa-ba-yon
**sablé** (Fr), sa-blé
**sablefish** (US), sa-ble-fish
**sabljarka** (SC), sab-ljar-ka
**sabores** (Pg), sa-bo-res
**sabzi** (Ia), sab-zi
**sabzi bora** (Ia), sab-zi bo-ra
**sabzi ké katlét** (Ia), sab-zi ké kat-lét
**sabzi kitchuri** (Ia), sab-zi kit-chu-ri
**sabzi kofta kalia** (Ia), sab-zi kof-ta ka-lia
**sacalait** (US), sac-a-lait
**sacarina** (Pg), sa-ca-ri-na
**saccharin** (US), sac-cha-rin
**saccoula** (Gk), sac-cou-la
**Sachertorte** (Gr), Sa-cher-tor-te
**sack** (GB), sack
**sacristain** (Fr), sa-cris-tain

**saddle** (US), sad-dle
**sadziki** (Gk), sa-dzi-ki
**safarjel** (Ar), sa-far-jel
**safflower oil** (US), saf-flow-er oil
**saffron** (US), saf-fron
**safran** (Fr), sa-fran
**Safran** (Gr), Sa-fran
**Saft** (GR), Saft
**Saftbraten** (Gr), Saft-bra-ten
**Saftig** (Gr), Saf-tig
**Saga Blue** (Da), Sa-ga Blue
**saganaki** (Gk), sa-ga-na-ki
**sage** (US), sage
**Sage Derby** (GB), Sage Der-by
**sago** (US), sa-go
**sago croquettes** (US), sa-go cro-quettes
**sagu** (In), sa-gu
**sahlab** (Ar), sah-lab
**Sahne** (Gr), Sah-ne
**Sahnekäse** (Gr), Sah-ne-kä-se
**Sahnemeerrettisch** (Gr), Sah-ne-meer-ret-tisch
**Sahnenkuchen** (Gr), Sah-nen-ku-chen
**saignant** (Fr), sai-gnant
**saigneux** (Fr), sai-gneux

**saim foogath** (Ia), saim foogath
**saimin** (Pl), sai-min
**saim rasa** (Ia), saim ra-sa
**Sainte Maure** (Fr), Sainte
Maure
**Saint-Germain, à la** (Fr),
Saint-Ger-main, à la
**Saint-Honoré** (Fr), Saint-Ho-
no-ré
**St. John's Bread** (US), St.
John's Bread
**Saint Otho** (Gr), Saint O-tho
**Saint Peter's fish** (US), Saint
Pe-ter's fish
**Saint-Pierre** (Fr), Saint-Pierre
**saisir** (Fr), sai-sir
**saith** (Sc), saith
**saithe** (US), saithe
**saive** (Ia), saive
**sajt** (Hu), sajt
**sajur** (In), sa-jur
**sakana no teriyaki** (Jp), sa-
ka-na no te-ri-ya-ki
**sakana ushojiru** (Jp), sa-ka-
na u-sho-ji-ru
**sakana yaki** (Jp), sa-ka-na ya-
ki
**sake** (Jp), sa-ke
**sake** (Jp), sa-ke
**sake no oyakomushi** (Jp), sa-
ke no o-ya-ko-mu-shi
**sake-zushi** (Jp), sa-ke-zu-shi
**sakhar** (Rs), sa-khar
**sakurambo** (Jp), sa-ku-ram-
bo
**sakuramochi** (Jp), sa-ku-ra-
mo-chi
**sal** (Sp, Pg), sal

**salaad** (Ia), sa-laad
**salaatinkastike** (Fi), sa-laat-
in-kas-ti-ke
**salaatti** (Fi), sa-laat-ti
**salad** (US), sal-ad
**salada** (Pg), sa-la-da
**salada mista** (Pg), sa-la-da
mis-ta
**salada praz** (Bu), sa-la-da
praz
**salada verde** (Pg), sa-la-da
ver-de
**salad burnet** (US), sal-ad
bur-net
**salad dressing** (US), sal-ad
dress-ing
**salade** (Fr), sa-lade
**salade de cresson** (Fr), sa-
lade de cres-son
**salade d'endive aux noix**
(Fr), sa-lade d'en-dive aux
noix
**salade d'épinards aux
champignons** (Fr), sa-lade
d'é-pi-nards aux cham-pi-
gnons
**salade de fonds d'arti-
chauts** (Fr), sa-lade de
fonds d'ar-ti-chauts
**salade de pissenlits** (Fr), sa-
lade de pis-sen-lits
**salade de saison** (Fr), sa-
lade de sai-son
**salade haricots verts** (Fr),
sa-lade ha-ri-cots verts
**salade mimosa** (Fr), sa-lade
mi-mo-sa
**salade niçoise** (Fr), sa-lade
ni-çoise

**salade panachée** (Fr), sa-lade pa-na-chée

**salade verte** (Fr), sa-lade verte

**salado** (Sp), sal-a-do

**salad oil** (US), sal-ad oil

**salaka** (Rs), sa-la-ka

**salam** (In), sa-lam

**salam** (Tr), sa-lam

**salame** (It), sa-la-me

**salame di fegato** (It), sa-la-me di fe-ga-to

**salamette** (It), sa-la-met-te

**salami** (It), sa-la-mi

**salammbô** (Fr), sa-lamm-bô

**salamoia** (It), sa-la-mo-ia

**salamon** (Ar), sa-la-mon

**salat** (Nw), sa-lat

**Salat** (Gr), Sa-lat

**salata** (Ar), sa-la-ta

**salata** (Tr), sa-la-ta

**salata** (Po), sa-la-ta

**salata** (Gk), sa-la-ta

**salataat** (Ar), sa-la-taat

**salatagurker** (Nw), sa-la-ta-gur-ker

**salata khadra** (Ar), sa-la-ta khad-ra

**salátaleves** (Hu), sa-lá-ta-le-ves

**salatalik** (Tr), sa-la-ta-lik

**salata melitzanes** (Gk), sa-la-ta me-lit-za-nes

**salata od morske ribe** (SC), sa-la-ta od mor-ske ri-be

**salata od piletine** (SC), sa-la-ta od pi-le-ti-ne

**salátaöntet** (Hu), sa-lá-ta-ön-tet

**salatet bataatis** (Ar), sa-la-tet ba-taa-tis

**salatet tamaatis** (Ar), sa-la-tet ta-maa-tis

**salatet tayheenah** (Ar), sa-la-tet tay-hee-nah

**salat og tomater** (Nw), sa-lat og to-ma-ter

**salátová marináda** (Cz), sa-lá-to-vá ma-ri-ná-da

**salát z rajských jablíček** (Cz), sa-lát z ra-j-ský-ch jab-lí-ček

**Salbei** (Gr), Sal-bei

**salça** (Tr), sal-ça

**salchichas con judias** (Sp), sal-chi-chas con ju-di-as

**salchichas en hojasd de maiz** (Sp), sal-chi-chas en ho-jasd de ma-iz

**salchichón** (Sp), sal-chi-chón

**salchisa** (Sp), sal-chi-sa

**salciccia** (It), sal-cic-cia

**salcissa** (It), sal-cis-sa

**sale** (It), sa-le

**salé** (Fr), sa-lé

**salep** (Tr), sa-lep

**salers fromage** (Fr), sa-lers fro-mage

**salgado** (Pg), sal-ga-do

**salicoque** (Fr), sa-li-coque

**Salisbury steak** (US), Salis-bu-ry steak

**salladsås** (Sw), sal-lad-sås

**Sally Lunn** (GB), Sal-ly Lunn

**Salm** (Gr), Salm

**salmagundi** (GB), sal-ma-gun-di

**salmagundi** (US), sal-ma-gun-di

**salmao** (Pg), sal-ma-o

**salmi** (Fr), sal-mi

**salmi de canard sauvage** (Fr), sal-mi de ca-nard sau-vage

**salmigondis** (Fr), sal-mi-gon-dis

**salmon** (US), salm-on

**salmon** (Sp), sal-mon

**salmonberry** (US), salm-on-ber-ry

**salmone** (It), sal-mo-ne

**salmone affumicato** (It), sal-mo-ne af-fu-mi-ca-to

**salmon en salsa verde** (Sp), sal-mon en sal-sa ver-de

**salmonete** (Pg), sal-mo-ne-te

**salmonetitos** (Sp), sal-mon-e-ti-tos

**salmon trout** (US), salm-on trout

**salmuera** (Sp), sal-mu-e-ra

**salpicão** (Pg), sal-pi-cã-o

**salpicon** (Fr), sal-pi-con

**salpicon de lengua** (Sp), sal-pi-con de len-gua

**salpicon de pescado** (Sp), sal-pi-con de pes-ca-do

**salsa** (It, Sp), sal-sa

**salsa** (Pg), sal-sa

**salsa** (Mx), sal-sa

**salsa alla milanese** (It), sal-sa al-la mi-la-ne-se

**salsa bianca** (It), sal-sa bi-an-ca

**salsa borracha** (Mx), sal-sa bor-ra-cha

**salsa casera** (Mx), sal-sa ca-ser-a

**salsa cruda** (Mx), sal-sa cru-da

**salsa d'acciughe** (It), sal-sa d'ac-ciu-ghe

**salsa d'aglio** (It), sal-sa d'a-gli-o

**salsa de chile guero** (Mx), sal-sa de chi-le gue-ro

**salsa de chile rojo** (Mx), sal-sa de chi-le ro-jo

**salsa di burro al gorgonzola** (It), sal-sa di bur-ro al gor-gon-zo-la

**salsa di carne** (It), sal-sa di car-ne

**salsa di funghi** (It), sal-sa di fun-ghi

**salsa di pignoli** (It), sal-sa di pi-gno-li

**salsa di pomodori** (It), sal-sa di po-mo-do-ri

**salsa inglesa** (Sp), sal-sa in-gle-sa

**salsa tonnata** (It), sal-sa ton-na-ta

**salsa verde** (It), sal-sa ver-de

**salsa verde** (Mx), sal-sa ver-de

**salsicce** (It), sal-sic-ce

**salsicha** (Pg), sal-si-cha

**salsifis frits** (Fr), sal-si-fis frits

**salsify** (US), sal-si-fy

**salt** (US), salt

**salt** (Nw, Sw), salt

**salteado** (Pg), sal-te-a-do

**saltfiskballer** (Nw), salt-fisk-ball-er

**salt-free** (US), salt-free

**saltimbocca** (It), sal-tim-boc-ca

**saltpeter** (US), salt-pe-ter

**salt pork** (US), salt pork

**salt-rising bread** (US), salt-ris-ing bread

**sàltsa** (Gk), sàl-tsa

**salty dog** (US), salt-y dog

**salva** (Pg), sal-va

**salvia** (It, Sp), sal-vi-a

**Salz** (Gr), Salz

**Salzbrühe** (Gr), Salz-brü-he

**Salzburger Mozart Kugel** (Gr), Salz-bur-ger Moz-art Ku-gel

**Salzburger Nockerl** (Gr), Salz-bur-ger Noc-kerl

**Salzgebäck** (Gr), Salz-ge-bäck

**Salzgurken** (Gr), Salz-gur-ken

**Salzkartoffeln** (Gr), Salz-kar-tof-feln

**samak** (Ar), sa-mak

**samaki** (Af-Swahili), sa-ma-ki

**samaki wa changu** (Af-Swahili), sa-ma-ki wa cha-ngu

**samak meshwi** (Ar), sa-mak mesh-wi

**samak moosa** (Ar), sa-mak moo-sa

**sambal** (In), sam-bal

**sambal goreng daging sapi** (In), sam-bal go-reng da-ging sa-pi

**sambal goreng udang** (In), sam-bal go-reng u-dang

**sambhar** (Ia), sam-bhar

**sambol** (In), sam-bol

**sambusa** (Af-Swahili), sa-mbu-sa

**sambusek** (Ar), sam-bu-sek

**samek** (Ar), sa-mek

**samfaina** (Sp), sam-fa-i-na

**şam fistiği** (Tr), şam fis-ti-ği

**samin** (Ar), sa-min

**samneh** (Ar), sam-neh

**samosa** (Ia), sa-mo-sa

**samp** (US), samp

**samphire** (US), sam-phire

**sampi** (Fi), sam-pi

**sämpylä** (Fi), säm-py-lä

**Samsøe** (Da), Sam-søe

**sand bakkelse** (Sw), sand bak-kel-se

**sand cake** (GB), sand cake

**sande** (Pg), san-de

**sandesh** (Ia), san-desh

**sandia** (Sp), san-di-a

**sandkage** (Da), sand-ka-ge

**Sandkuchen** (Gr), Sand-ku-chen

**sand pear** (US), sand pear

**sandra** (Fr), san-dra

**sandwich** (GB), sand-wich

**sandwich** (US), sand-wich

**sang, au** (Fr), sang, au

**sangaree** (US), san-ga-ree

**sang chu** (Kr), sang chu

**sanglier** (Fr), sang-li-er

**sangrento** (Pg), san-gren-to

**sangria** (Sp), san-gri-a

**sangrita** (Sp), san-gri-ta

**sangsun hwe** (Kr), sang-sun hwe

**sangue** (It), san-gue

**sanguinaccio** (It), san-gui-na-ccio

**San Pedro** (Sp), San Pe-dro
*sansai* (Jp), san-sai
*sansho* (Jp), san-sho
*santaka* (Jp), san-ta-ka
*santan* (In), san-tan
*santaraa* (Ia), san-ta-raa
*santen* (In), san-ten
*santola* (Pg), san-to-la
*santola gratinada* (Pg), san-to-la gra-ti-na-da
*saos* (In), saos
*sap* (Du), sap
*sapodilla* (US), sap-o-dil-la
*saporoso* (It), sa-po-ro-so
**Sapsago** (Gr), Sap-sa-go
*sapsis* (US), sap-sis
*sapucaya nut* (Pg), sap-u-cay-a nut
**Saracen corn** (US), Sar-a-cen corn
*sarada yu* (Jp), sa-ra-da yu
**Sarah Bernhardt cookie** (US), Sar-ah Bern-hardt cook-ie
*sarang burang* (In), sa-rang bu-rang
*şarap* (Tr), şa-rap
*sarapatel* (Pg), sa-ra-pa-tel
**Saratoga chips** (US), Sar-a-to-ga chips
**Saratoga chops** (US), Sar-a-to-ga chops
*sarcelle* (Fr), sar-celle
*sard* (Fr), sard
*sarda* (Pg), sar-da
*sardalya* (Tr), sar-dal-ya
*sardalya tavasi* (Tr), sar-dal-ya ta-va-si
*sardeen* (Ar), sar-deen
**Sardellen** (Gr), Sar-del-len

*sarden* (In), sar-den
*sardheles* (Gk), sar-dhe-les
*sardiini* (Fi), sar-dii-ni
*sardin* (Sw), sar-din
*sardine* (It), sar-di-ne
**Sardinen** (Gr), Sar-di-nen
*sardiner* (Nw), sar-din-er
*sardines* (US), sar-dines
*sardinha* (Pg), sar-di-nha
*sardinhas de caldeirada* (Pg), sar-di-nhas de cal-de-i-ra-da
**Sardo Romano** (It), Sar-do Ro-ma-no
*sardynka* (Po), sar-dyn-ka
*sárgabarack* (Hu), sár-ga-ba-rack
*sárgadinnye* (Hu), sár-ga-din-nye
*sárgarépa* (Hu), sár-ga-ré-pa
*šargarepe* (SC), šar-ga-re-pe
*sarma* (SC), sar-ma
*sarmi* (Bu), sar-mi
*sāroin* (Jp), sā-roin
*sarriette* (Fr), sar-ri-ette
*sarsaparilla* (Sp), sar-sa-pa-ri-lla
*sarson kaa saag* (Ia), sar-son kaa saag
*sås* (Sw), sås
*sashimi* (Jp), sa-shi-mi
*sassafras* (US), sas-sa-fras
*sassafras mead* (US), sas-sa-fras mead
*sassefrica* (It), sas-se-fri-ca
*satay* (In), sa-tay
*saté* (In), sa-té
*saté ajam* (In), sa-té a-jam
*satō* (Jp), sa-tō

*satoimo* (Jp), sa-to-i-mo

*satsuma* (US), sat-su-ma

*satsuma age* (Jp), sat-su-ma age

*satsuma imo* (Jp), sat-su-ma i-mo

*saturated fat* (US), sat-u-rat-ed fat

*saturated fatty acid* (US), sat-u-rat-ed fat-ty ac-id

*Saubohnen* (Gr), Sau-boh-nen

*sauce* (US), sauce

*sauce piquante de soya* (Fr), sauce pi-quante de so-ya

*saucijsjes* (Du), sau-cij-sjes

*saucijzenbroodje* (Du), sau-cij-zen-brood-je

*saucisse* (Fr), sau-cisse

*saucisses au vin blanc* (Fr), sau-cis-ses au vin blanc

*saucisson* (Fr), sau-cis-son

*saucisson à l'aioli* (Fr), sau-cis-son à l'ai-o-li

*Sauerampfer* (Gr), Sau-er-amp-fer

*Sauerbraten* (Gr), Sau-er-bra-ten

*Sauerkraut* (Gr), Sau-er-kraut

*Sauerkraut mit Schweinebauch* (Gr), Sau-er-kraut mit Schwei-ne-bauch

*Sauermilchkäse* (Gr), Sau-er-milch-kä-se

*sauge* (Fr), sauge

*saumon* (Fr), sau-mon

*saumon d'Ecosse fumé* (Fr), sau-mon d'E-cosse fu-mé

*saumon en gelée* (Fr), sau-mon en ge-lée

*saumon fumé* (Fr), sau-mon fu-mé

*saumon glacé* (Fr), sau-mon gla-cé

*saumon poché* (Fr), sau-mon po-ché

*saumure* (Fr), sau-mure

*saunamakkara* (Fi), sa-u-na-mak-ka-ra

*saunf* (Ia), saunf

*Saure Rahmsauce* (Gr), Sau-re Rahm-sau-ce

*Saure Sahne* (Gr), Sau-re Sah-ne

*sauro* (It), sau-ro

*saus* (Nw), saus

*saus* (Du, In), saus

*sausage* (US), sau-sage

*saussiski v tomate* (Rs), saus-si-ski v to-ma-te

*sauté* (US, Fr), sau-té

*sautérat* (Sw), sau-té-rat

*sautiert* (Gr), sau-tiert

*sauvage* (Fr), sau-vage

*savanyú káposzta* (Hu), sa-va-nyú ká-posz-ta

*savarin* (Fr), sa-va-rin

*sável* (Pg), sá-vel

*savoiardi* (It), sa-voi-ar-di

*savory* (US), sa-vor-y

*savory* (US), sa-vor-y

*savouries* (GB), sa-vour-ies

*savoyarde* (Fr), sa-voy-arde

*savoy biscuits* (GB), sa-voy bis-cuits

*savoy cabbage* (Fr), sa-voy cab-bage

*savu* (Fi), sa-vu

*savukala* (Fi), sa-vu-ka-la

*savukalasalaatti* (Fi), sa-vu-ka-la-sa-laat-ti

*savusiika* (Fi), sa-vu-si-i-ka

*savusilakka* (Fi), sa-vu-si-lak-ka

*sawi* (In), sa-wi

*sawo* (In), sa-wo

*saya éndo* (Jp), sa-ya én-do

*sayoo chatni* (Ia), sa-yoo chat-ni

*sayur* (In), sa-yur

*sayur kuning* (In), sa-yur ku-ning

*sazené v mléku* (Cz), sa-ze-né v mlé-ku

*sbanikh* (Ar), sba-nikh

*sbrinz* (It), sbrinz

*sbrisolona* (It), sbri-so-lo-na

*scadán* (Ir), scad-án

*scald* (US), scald

*scallions* (US), scal-lions

*scallop, to* (US), scal-lop, to

*scalloped veal* (US), scal-loped veal

*scallops* (US), scal-lops

*scaloppine* (It), sca-lop-pi-ne

*scaloppine al funghi* (It), sca-lop-pi-ne al fun-ghi

*scaloppine al sedano* (It), sca-lop-pi-ne al se-da-no

*Scamorze* (It), Sca-mor-ze

*scampi* (It), scam-pi

*scampi* (US), scam-pi

*scampi fra diavole* (It), scam-pi fra dia-vo-le

*Scanno* (It), Scan-no

*scarola* (It), sca-ro-la

*scarpariello, alla* (It), scar-pa-riel-lo, al-la

*scarzetta* (It), scar-zet-ta

*scate* (Sc), scate

*Schabzieger* (Gr), Schab-zie-ger

*Schalotten* (Gr), Scha-lot-ten

*Schaltiere* (Gr), Schal-tie-re

*schapevlees* (Du), scha-pe-vlees

*scharf* (Gr), scharf

*Schaschlik* (Gr), Schasch-lik

*Schaum* (Gr), Schaum

*Schaumrollen* (Gr), Schaum-rol-len

*Schaumwein* (Gr), Schaum-wein

*Schellfisch* (Gr), Schell-fisch

*schelvis* (Du), schel-vis

*schelviskuitjes* (Du), schel-vis-kui-tjes

*schelvislever* (Du), schel-vis-le-ver

*schiacciata* (It), schia-ccia-ta

*Schildkrötensuppe* (Gr), Schild-krö-ten-sup-pe

*Schinken* (Gr), Schin-ken

*Schinkenfleckerl* (Gr), Schin-ken-flec-kerl

*Schinkenomelette* (Gr), Schin-ken-om-e-let-te

*Schlachtplatte* (Gr), Schlacht-plat-te

*Schlag* (Gr), Schlag

*Schlagobers* (Gr-Austria), Schla-go-bers

*Schlagsahne* (Gr), Schlag-sah-ne

*Schlegel* (Gr), Schle-gel

*Schlesisches Himmelreich* (Gr), Schles-isch-es Him-mel-reich

**Schlosser Buben** (Gr-Austria), Schlos-ser Bu-ben

**Schmalz** (Gr), Schmalz

**Schmalzgebackenes** (Gr), Schmalz-ge-ba-cken-es

**Schmandschinken** (Gr), Schmand-schin-ken

**Schmarrn** (Gr), Schmarrn

**Schmierkäse** (Gr), Schmier-kä-se

**Schmorfleisch** (Gr), Schmor-fleisch

**Schnäpel** (Gr), Schnä-pel

**Schnaps** (Gr), Schnaps

**Schnecken** (Gr), Schne-cken

**Schneenockerln** (Gr), Schnee-noc-kerln

**Schnittbohne** (Gr), Schnitt-boh-ne

**Schnitte** (Gr), Schnit-te

**Schnittlauch** (Gr), Schnitt-lauch

**Schnitz** (Gr), Schnitz

**Schnitzel** (Gr), Schnit-zel

**Schnitzel Holstein** (Gr), Schnit-zel Hol-stein

**Schnitz und Gnepp** (Gr), Schnitz und Gnepp

**Schnupfnudel** (Gr), Schnupf-nu-del

**Schokolade** (Gr), Scho-ko-la-de

**Schokoladeneis** (Gr), Scho-ko-la-den-eis

**schol** (Du), schol

**Scholle** (Gr), Schol-le

**Schöpsenschlegel** (Gr), Schöp-sen-schle-gel

**Schotensuppe** (Gr), Scho-ten-sup-pe

**Schrotbrot** (Gr), Schrot-brot

**Schüblig** (Gr), Schü-blig

**schuimpjes** (Du), schuimp-jes

**Schulter** (Gr), Schul-ter

**Schupfnudeln** (Gr), Schupf-nu-deln

**Schutzenkäse** (Gr), Schut-zen-kä-se

**Schwämme** (Gr-Austria), Schwäm-me

**Schwärtelbraten** (Gr), Schwär-tel-bra-ten

**Schwarzbrot** (Gr), Schwarz-brot

**Schwarzer Kaffee** (Gr), Schwar-zer Kaf-fee

**Schwarzfisch** (Gr), Schwarz-fisch

**Schwarzsauer** (Gr), Schwarz-sau-er

**Schwarzwälder Kirschtorte** (Gr), Schwarz-wäl-der Kirsch-tor-te

**Schwarzwürste** (Gr), Schwarz-würs-te

**Schwarzwurzeln** (Gr), Schwarz-wur-zeln

**Schweinebauch** (Gr), Schwei-ne-bauch

**Schweinebraten mit einer Kruste** (Gr), Schwei-ne-bra-ten mit ei-ner Krus-te

**Schweinebrust mit Äpfeln** (Gr), Schwei-ne-brust mit Äp-feln

**Schweinefleisch** (Gr), Schwei-ne-fleisch

*Schweinefleisch im bier* (Gr), Schwei-ne-fleisch im bier
*Schweinekeule* (Gr), Schwei-ne-keu-le
*Schweinekotelett* (Gr), Schwei-ne-ko-te-lett
*Schweineohren* (Gr), Schwei-ne-oh-ren
*Schweinepfeffer* (Gr), Schwei-ne-pfef-fer
*Schweinerippchen* (Gr), Schwei-ne-ripp-chen
*Schweinerucken* (Gr), Schwei-ne-ru-cken
*Schweineschenkel* (Gr), Schwei-ne-schen-kel
*Schweinsfilets mit Saure Sahne* (Gr), Schweins-fil-ets mit Sau-re Sah-ne
*Schweinsjungfernbraten* (Gr), Schweins-jung-fern-bra-ten
*Schweinskarre* (Gr), Schweins-kar-re
*Schweinssulz* (Gr-Austria), Schweins-sulz
*Schweizerkäse* (Gr), Schwei-zer-kä-se
*Schwertfisch* (Gr), Schwert-fisch
*sciroppo* (It), sci-rop-po
*scone* (Sc), scone
*score* (US), score
*scorpion fish* (US), scor-pi-on fish
*scorzonera* (Sp), scor-zo-ner-a
*Scotch broth* (Sc), Scotch broth
*Scotch eggs* (Sc), Scotch eggs

*Scotch tender* (US), Scotch ten-der
*Scotch woodcook* (Sc), Scotch wood-cook
*scottadito* (It), scot-ta-di-to
*scramble* (US), scram-ble
*scrapple* (Gr), scrap-ple
*scripture cake* (US), scrip-ture cake
*scrod* (US), scrod
*sculpin* (US), scul-pin
*scungilli* (It), scun-gil-li
*scup* (US), scup
*scuppernong* (US), scup-per-nong
*sea bass* (US), sea bass
*sea bean* (GB), sea bean
*sea biscuit* (US), sea bis-cuit
*sea-bob* (US), sea-bob
*sea bream* (US), sea bream
*sea cucumber* (US), sea cu-cum-ber
*sea dates* (It), sea dates
*sea ear* (US), sea ear
*sea fennel* (US), sea fen-nel
*seafood* (US), sea-food
*sea kale* (US), sea kale
*sea moss* (US), sea moss
*sea pie* (US), sea pie
*sear* (US), sear
*sea salt* (US), sea salt
*sea slug* (US), sea slug
*seasnails* (US), sea-snails
*seasoned salts* (US), sea-soned salts
*seasoning* (US), sea-son-ing
*sea swallow* (GB), sea swal-low
*sea tongue* (US), sea tongue
*sea trout* (US), sea trout

*sea urchin* (US), sea ur-chin
*seaweed* (US), sea-weed
*seaweed bread* (US), sea-weed bread
*seb chatni* (Ia), seb chat-ni
*sebze* (Tr), seb-ze
*sebze çorbasi* (Tr), seb-ze çor-ba-si
*sec* (Fr), sec
*šećer* (SC), še-ćer
*séché* (Fr), sé-ché
*séco* (Pg), sé-co
*sedano* (It), se-da-no
*sedano-rapa* (It), se-da-no-ra-pa
*seeg* (Rs), seeg
*seekh kabab* (Ia), seekh ka-bab
*Seekrabben* (Gr), See-krab-ben
*Seelachs* (Gr), See-lachs
*See-ohr* (Gr), See-ohr
*seesehr* (Tr), see-sehr
*seesehri aghtsan* (Tr), see-sehri agh-tsan
*Seezunge* (Gr), See-zun-ge
*sefrina* (Ar), se-fri-na
*şeftali* (Tr), şef-ta-li
*segala* (It), se-ga-la
*sehr* (Tr), sehr
*şehriye çorbasi* (Tr), şeh-ri-ye çor-ba-si
*seigle* (Fr), seigle
*sekahedelmakeitto* (Fi), se-ka-he-del-ma-ke-it-to
*sekahedelmakiisseli* (Fi), se-ka-he-del-ma-ki-is-se-li
*sekaná pečeně* (Cz), se-ka-ná pe-če-ně
*sekané* (Cz), se-ka-né

*şeker* (Tr), şe-ker
*sekihan* (Jp), se-ki-han
*seksu* (Ar), sek-su
*Sekt* (Gr), Sekt
*sel* (Fr), sel
*sel'd'* (Rs), sel'd'
*selderie* (Du), sel-de-rie
*seleek* (Ar), se-leek
*self-rising flour* (US), self-ris-ing flour
*selha chawal* (Ia), se-lha cha-wal
*selino avgolemono* (Gk), se-li-no av-go-le-mo-no
*selinon me ladi* (Gk), se-li-non me la-di
*Selkirk bannock* (Sc), Sel-kirk ban-nock
*selle* (Fr), selle
*selle d'agneau* (Fr), selle d'agn-eau
*selle de pre-salé desosée* (Fr), selle de pre-sa-lé de-so-sée
*selleri* (Nw), sell-e-ri
*selleri* (Fi), sel-le-ri
*Sellerie* (Gr), Sel-le-rie
*Selleriesalat* (Gr), Sel-lerie-sa-lat
*Selles-sur-Cher* (Fr), Selles-sur-Cher
*selodka* (Rs), sel-od-ka
*selters* (Nw), sel-ters
*seltzer* (US), selt-zer
*selvaggina* (It), sel-vag-gi-na
*sém* (Ia), sém
*sembei* (Jp), sem-bei
*sëmga* (Rs), sëm-ga
*semi di melone* (It), se-mi di me-lo-ne

*semifreddo* (It), se-mi-fred-do

*semifrio* (Sp), se-mi-fri-o

*semillas* (Sp), sem-illas

*semillas tostados de cala-baza* (Sp), sem-illas tos-ta-dos de ca-la-ba-za

*semimoist foods* (US), sem-i-moist foods

*semisweet chocolate* (US), sem-i-sweet choc-o-late

*semlor* (Sw), sem-lor

*semmel* (US), sem-mel

*Semmelklosse* (Gr), Sem-mel-klos-se

*Semmel knodel* (Gr), Sem-mel kno-del

*semolina* (US), sem-o-li-na

*semur* (In), se-mur

*semur ayam* (In), se-mur a-yam

*semur lidah* (In), se-mur li-dah

*semur terong* (In), se-mur te-rong

*senap* (Sw), se-nap

*senape* (It), se-na-pe

*senf* (SC), senf

*Senf* (Gr), Senf

*Senfgurken* (Gr), Senf-gur-ken

*sennep* (Da, Nw), sen-nep

*sepia* (Sp), sep-ia

*seppie* (It), sep-pie

*seppie al pomodoro* (It), sep-pie al po-mo-do-ro

*sequestrant* (US), se-ques-trant

*ser* (Po), ser

*šerbet* (SC), šer-bet

*şerbet* (Tr), şer-bet

*sereh* (In), se-reh

*serendipity berry* (US), ser-en-dip-i-ty ber-ry

*sergevil* (Tr), ser-ge-vil

*sergevili anoosh* (Tr), ser-ge-vi-li a-noosh

*sergevili bastegh* (Tr), ser-ge-vi-li bas-tegh

*sergevili osharag* (Tr), ser-ge-vi-li o-sha-rag

*seri* (Jp), se-ri

*seroendeng* (In), seroen-deng

*serpenyös rostelyos* (Hu), ser-pe-nyös ros-te-lyos

*Serrano chile* (Mx), Ser-ra-no chi-le

*serrucho en escabeche* (Sp), ser-ru-cho en es-ca-be-che

*sertésborda parasztosan* (Hu), ser-tés-bor-da pa-rasz-to-san

*serundeng* (In), serun-deng

*serviette, à la* (Fr), ser-vi-ette, à la

*Serviettenklöss mit Birnen und Bohnen* (Gr), Ser-vi-et-ten-klöss mit Bir-nen und Boh-nen

*seryi skat* (Rs), ser-yi skat

*sesame oil* (US), ses-a-me oil

*sesame seeds* (US), ses-a-me seeds

*se smetanou* (Cz), se sme-ta-no-u

*sesos de ternera* (Sp), se-sos de ter-ne-ra

*set, to* (US), set, to

*setas* (Sp), se-tas

**setas sobre las parillas** (Sp), se-tas so-bre las pa-ri-llas

**setrup** (In), set-rup

**sevaee** (Ia), se-vaee

**seviche** (Sp), se-vi-che

**seviche de vieiras** (Sp), se-vi-che de vi-e-ir-as

**Seville orange** (US), Se-ville orange

**sevruga** (Rs), sev-ru-ga

**sfarjal** (Ar), sfar-jal

**sfeeha** (Ar), sfee-ha

**sfingi** (It), sfin-gi

**sfogato** (Gk), sfo-ga-to

**sfogliatelle** (It), sfo-gli-a-tel-le

**sfogliatine di crema** (It), sfo-gli-a-ti-ne di cre-ma

**sformato** (It), sfor-ma-to

**sformoto di tonno** (It), sfor-mo-to di ton-no

**sgombro** (It), sgom-bro

**shābetto** (Jp), shā-bet-to

**shabu-shabu** (Jp), sha-bu-sha-bu

**shad** (US), shad

**shaddock** (US), shad-dock

**shad roe** (US), shad roe

**shahbahr** (Ar), shah-bahr

**shakarkand** (Ia), sha-kar-kand

**shake** (Jp), shake

**shakuwlaata** (Ar), sha-kuw-laa-ta

**shā-là jyàng** (Ch), shā-là jy-àng

**shaljam** (Ia), shal-jam

**shaljam bharta** (Ia), shal-jam bhar-ta

**shaljam rasa** (Ia), shal-jam ra-sa

**shallot** (US), shal-lot

**shammehm** (Ar), sham-mehm

**shamme kabab** (Ia), sham-me ka-bab

**shammoama** (Ar), sham-moa-ma

**shamouti** (US), sha-mou-ti

**shàn-bèi-ké** (Ch), shàn-bèi-ké

**shandy** (GB), shan-dy

**shandygaff** (GB), shan-dy-gaff

**shank** (US), shank

**shao** (Ch), shao

**shao mài** (Ch), shao mài

**sharbat** (Ia), shar-bat

**sharbati** (Af-Swahili), shar-ba-ti

**shareeyee** (Ar), sha-ree-yee

**shark** (US), shark

**shark fin** (US), shark fin

**shashlik** (Rs), sha-shlik

**shataavar** (Ia), sha-taa-var

**shawirma** (Ar), sha-wir-ma

**shchi** (Rs), shchi

**she-crab soup** (US), she-crab soup

**sheepshead** (US), sheeps-head

**shehrieh** (Tr), sheh-ri-eh

**sheldrake** (GB), shel-drake

**shelisheli** (Af-Swahili), she-li-she-li

**shell bean** (US), shell bean

**shellfish** (US), shell-fish

**shell steak** (US), shell steak

*shēng tsài* (Ch), shēng tsài

*shepherd's pie* (GB), shepherd's pie

*shepherd's purse* (US), shepherd's purse

*sherbet* (US), sher-bet

*sheriya miftoon* (Ar), she-ri-ya mif-toon

*sherry* (Sp), sher-ry

*shé-tóu* (Ch), shé-tóu

*shī* (Ar), shī

*shi bilabun* (Ar), shi bi-la-bun

*shi binayna* (Ar), shi bi-nay-na

*shichimi* (Jp), shi-chi-mi

*shiitake* (Jp), shi-i-ta-ke

*shikaar kaa gosht* (Ia), shi-kaar kaa gosht

*shī metallig* (Ar), shī me-tal-lig

*shinmei* (Jp), shin-mei

*shio* (Jp), shi-o

*shio sembei* (Jp), shi-o sem-bei

*shioyaki* (Jp), shi-o-ya-ki

*shiozuki* (Jp), shi-o-zu-ki

*ship biscuit* (US), ship biscuit

*ship caviar* (US), ship ca-vi-ar

*shiraita kombu* (Jp), shi-ra-i-ta kom-bu

*shirataki* (Jp), shi-ra-ta-ki

*shiratamako* (Jp), shi-ra-ta-ma-ko

*shiriyyi* (Ar), shi-riy-yi

*shiro* (Jp), shi-ro

*shiro miso* (Jp), shi-ro mi-so

*shirona* (Jp), shi-ro-na

*shirred eggs* (US), shirred eggs

*shirring* (US), shir-ring

*shiruko* (Jp), shi-ru-ko

*shirumiso* (Jp), shi-ru-mi-so

*shirumono* (Jp), shi-ru-mo-no

*shish kabob* (US), shish ka-bob

*shiso* (Jp), shi-so

*shoat* (US), shoat

*shoga* (Jp), sho-ga

*shoga sembei* (Jp), sho-ga sem-bei

*shokishoki* (Af-Swahili), sho-ki-sho-ki

*shomin* (Tr), sho-min

*shomini aboor* (Tr), sho-mini a-boor

*shoofly pie* (US), shoo-fly pie

*shooshma* (Tr), shoosh-ma

*shooshmayov hahts* (Tr), shoosh-ma-yov hahts

*shooshmayov pleet* (Tr), shoosh-ma-yov pleet

*shorba* (Ar, Ia), shor-ba

*shorbet ads* (Ar), shor-bet ads

*shorbet basal* (Ar), shor-bet ba-sal

*shorbet ferakh* (Ar), shor-bet fe-rakh

*shorbet khudahr* (Ar), shor-bet khu-dahr

*shorbet lahmah* (Ar), shor-bet lah-mah

*shorbet samak* (Ar), shor-bet sa-mak

*shorbet shreeya* (Ar), shor-bet shree-ya

**shore dinner** (US), shore din-ner

**shortbread** (Sc), short-bread

**shortcake** (US), shortcake

**shortening** (US), short-en-ing

**short-grain rice** (US), short-grain rice

**short loin** (US), short loin

**shortnin' bread** (US), short-nin' bread

**short ribs** (US), short ribs

**shorva** (Ia), shor-va

**shoulder** (US), should-er

**shoulder of pork** (US), should-er of pork

**shoyu** (Jp), sho-yu

**shproti** (Rs), shpro-ti

**shr-dz** (Ch), shr-dz

**shred** (US), shred

**Shrewsbury cake** (US), Shrews-bu-ry cake

**shrimp** (US), shrimp

**shrimp paste** (US), shrimp paste

**shuànyángròu** (Ch), shu-àn-yáng-ròu

**shuck** (US), shuck

**shukti** (Ia), shuk-ti

**shukto** (Ia), shuk-to

**shummam** (Ar), shum-mam

**shungiku** (Jp), shun-gi-ku

**siadle mleko** (Po), siad-le mle-ko

**sianliha** (Fi), si-an-li-ha

**sids** (Sc), sids

**sieniä** (Fi), si-e-ni-ä

**sienimuhennos** (Fi), si-e-ni-mu-hen-nos

**sienisalaatti** (Fi), si-e-ni-sa-laat-ti

**siero di latte** (It), sie-ro di lat-te

**sieve** (US), sieve

**sift** (US), sift

**sig** (Rs), sig

**sigara böreği** (Tr), si-ga-ra bö-re-ği

**siğir** (Tr), siğir

**sigtebrød** (Da), sig-te-brød

**siika** (Fi), sii-ka

**sik** (Sw), sik

**sikampouri kabab** (Ia), si-kam-pou-ri ka-bab

**siki** (Af-Swahili), si-ki

**sikotakia tiganata** (Gk), si-ko-ta-kia ti-ga-na-ta

**silakka** (Fi), si-lak-ka

**silakkalaatikko** (Fi), si-lak-ka-laa-tik-ko

**silakkapihvit** (Fi), si-lak-ka-pih-vit

**silakkarullat** (Fi), si-lak-ka-rul-lat

**silcock** (Sc), sil-cock

**sild** (Nw), sild

**sild** (Da), sild

**sildeboller** (Nw), sil-de-boll-er

**sildeflyndere** (Da), sil-de-flyn-de-re

**sildegryn** (Nw), sil-de-gryn

**sildesalat** (Nw), sil-de-sa-lat

**sill** (Sw), sill

**sillbullar** (Sw), sill-bul-lar

**sillgratäng** (Sw), sill-gra-täng

**silli** (Fi), sil-li

**sillisalaatti** (Fi), sil-li-sa-laat-ti

**sillock** (Sc), sill-ock

*sill polsa* (Sw), sill pol-sa
*sillsalat* (Sw), sill-sa-lat
*silq* (Ar), silq
*silvano* (It), sil-va-no
*silver and gold leaf* (Ia), sil-ver and gold leaf
*silverside* (GB), sil-ver-side
*silverside* (GB), sil-ver-side
*sima* (Fi), si-ma
*simit* (Tr), si-mit
*simlaa mirch* (Ia), sim-laa mirch
*simmaq* (Ar), sim-maq
*simmehm* (Ar), sim-mehm
*simmer* (US), sim-mer
*simnel* (GB), sim-nel
*simpukat* (Fi), sim-pu-kat
*simsim* (Ar), sim-sim
*sinaasappel* (Du), si-naas-ap-pel
*sinaasappelsap* (Du), si-naas-ap-pel-sap
*sinappi* (Fi), si-nap-pi
*sinappisilakka* (Fi), si-nap-pi-si-lak-ka
*singara* (Ia), sin-ga-ra
*singaree* (US), sin-ga-ree
*singe* (US), singe
*sini kufteh* (Tr), si-ni kuf-teh
*sinsullo* (Kr), sin-sul-lo
*sippee* (Ia), sip-pee
*sippets* (US), sip-pets
*sipuli* (Fi), si-pu-li
*sipulikeitto* (Fi), si-pu-li-ke-it-to
*sipulipihvi* (Fi), si-pu-li-pih-vi
*sir* (SC), sir
*sirće* (SC), sir-će
*sirke* (Tr), sir-ke
*sirloin* (US), sir-loin

*sirloin chops* (US), sir-loin chops
*sirloin roast* (US), sir-loin roast
*sirloin tip* (US), sir-loin tip
*sirnaya* (Rs), sir-na-ya
*sirniki* (Rs), sir-ni-ki
*sirop* (Fr), si-rop
*şişe suyu* (Tr), şi-şe su-yu
*şiş kebab* (Tr), şiş ke-bab
*sitron* (Nw), si-tron
*sitronfromasje* (Nw), si-tron-fro-mas-je
*sitruuna* (Fi), sit-ruu-na
*sitruunakohokas* (Fi), sit-ruu-na-ko-ho-kas
*siyah havyar* (Tr), si-yah hav-yar
*siyah şarap* (Tr), si-yah şa-rap
*sjelé* (Nw), sje-lé
*sjokolade* (Nw), sjo-ko-la-de
*sjömansbiff* (Sw), sjö-mans-biff
*sjøørret* (Nw), sjø-ør-ret
*skaldjurssallad* (Sw), skal-djurs-sal-lad
*skaldjursstuvning* (Sw), skal-djurs-stuv-ning
*skansko potatis* (Sw), skan-sko po-ta-tis
*skärbönor* (Sw), skär-bö-nor
*skarpsås* (Sw), skarp-sås
*skarpsill* (Sw), skarp-sill
*skate* (US), skate
*skembe* (Gk), skem-be
*škembići* (SC), škem-bi-ći
*skewer, to* (US), skew-er, to
*skim* (US), skim
*skim milk* (US), skim milk
*skinka* (Sw), skin-ka

*skinkbullar* (Sw), skink-bul-lar

*skinke* (Da), skin-ke

*skinke med røræg* (Da), skin-ke med rør-æg

*skinke og egg* (Nw), skink-e og egg

*skinklada* (Sw), skink-la-da

*skirlie* (Sc), skir-lie

*skirret* (US), skir-ret

*skirt steak* (US), skirt steak

*sköldpadda* (Sw), sköld-pad-da

*sköldpaddssoppa* (Sw), sköld-padds-sop-pa

*školjke* (SC), školj-ke

*skopové maso* (Cz), sko-po-vé ma-so

*skordalia* (Gk), skor-da-lia

*skordalia me pignolia* (Gk), skor-da-lia me pig-no-lia

*skorpor* (Sw), skor-por

*skoumbria* (Gk), skoum-bri-a

*skummet melk* (Nw), skumm-et melk

*sla* (Du), sla

*slab bacon* (US), slab ba-con

*sladké* (Cz), slad-ké

*sladoled od vanilije* (SC), sla-do-led od va-ni-li-je

*slagroom* (Du), slag-room

*slaked lime* (US), slaked lime

*slanina* (SC), sla-ni-na

*slapjack* (US), slap-jack

*sla saus* (Du), sla saus

*slatrokka* (Sw), slat-rok-ka

*slaw* (GB), slaw

*sled* (Cz), sled

*sledz* (Po), sledz

*sledz marynowany ze smie-tanom* (Po), sledz ma-ry-no-wa-ny ze smie-ta-nom

*šlebačka* (Cz), šle-hač-ka

*slepičí polévka* (Cz), sle-pi-čí po-lév-ka

*slethvarre* (Da), sleth-var-re

*slettvar* (Nw), slett-var

*slice* (US), slice

*sling* (GB), sling

*sliver* (US), sli-ver

*slivovitz* (SC), sli-vo-vitz

*šljive* (SC), šlji-ve

*slodkie* (Po), slod-kie

*sloe* (GB), sloe

*sloke* (GB), sloke

*sloppy joe* (US), slop-py joe

*slotssteg* (Da), slots-steg

*slottsstek* (Sw), slotts-stek

*slumgullion* (US), slum-gul-lion

*slump* (US), slump

*småbröd* (Sw), små-bröd

*småfranska* (Sw), små-fran-ska

*småkage* (Da), små-ka-ge

*småkaker* (Nw), små-kak-er

*smaker* (Sw), sma-ker

*små köttbullar* (Sw), små kött-bul-lar

*småländsk ostkaka* (Sw), små-länd-sk ost-ka-ka

*smallage* (US), small-age

*småltsill* (Sw), smålt-sill

*smasill* (Sw), sma-sill

*småvarmt* (Sw), små-varmt

*smažená vejce* (Cz), sma-že-ná vej-ce

*smažené* (Cz), sma-že-né

*smelt* (US), smelt

**smen** (Ar), smen
**smetana** (Rs), sme-ta-na
**smetanick** (Rs), sme-ta-nick
**smid** (Ar), smid
**śmietanka** (Po), śmie-tan-ka
**smitane** (Fr), smi-tane
**Smithfield ham** (US), Smithfield ham
**smoke** (US), smoke
**smoked butt** (US), smoked
butt
**smokies** (US), smok-ies
**smokve** (SC), smok-ve
**smör** (Sw), smör
**smør** (Da, Nw), smør
**smørbrød** (Nw), smør-brød
**smörgåsar** (Sw), smör-gå-sar
**smörgåsbord** (Sw), smör-
gås-bord
**smørkage** (Da), smør-ka-ge
**smørrebrød** (Da), smør-re-
brød
**smult** (Nw), smult
**smultron** (Sw), smul-tron
**smyrnaika** (Gk), smyr-na-i-
ka
**snails** (US), snails
**snap bean** (US), snap bean
**sneeuwballen** (Du), snee-
uw-bal-len
**snegelhus** (Nw), sne-gel-hus
**snijbonen** (Du), snij-bo-nen
**snipe** (GB), snipe
**snittbønner** (Nw), snitt-bø-
nner
**snitter** (Da), snit-ter
**snoek** (Du), snoek
**snöripa** (Sw), snö-ri-pa
**snow eggs** (US), snow eggs
**snow peas** (US), snow peas

**so** (SC), so
**só** (Hu), só
**soak, to** (US), soak, to
**soba** (Jp), so-ba
**socker** (Sw), so-cker
**socivo salata** (SC), so-ci-vo
sa-la-ta
**soda** (US), so-da
**soda bread** (US), so-da bread
**soda water** (US), so-da wa-ter
**sodium-free** (US), so-di-um-
free
**šodó** (Cz), šo-dó
**sødsuppe** (Da), sød-sup-pe
**soepen** (Du), soep-en
**soffritto** (It), sof-frit-to
**sofrito** (Sp), so-fri-to
**soft-serve** (US), soft-serve
**soğan** (Tr), so-ğan
**sogliola** (It), so-glio-la
**sogliole alla marinara** (It),
so-glio-le al-la ma-ri-na-ra
**soğuk et** (Tr), so-ğuk et
**sohk** (Rs), sohk
**sohl** (Rs), sohl
**Soissons** (Fr), Sois-sons
**sokeri** (Fi), so-ke-ri
**sokh** (Tr), sokh
**sokhov tutoom** (Tr), sokh-ov
tu-toom
**sok od narandže** (SC), sok
od na-ran-dže
**sól** (Po), sól
**sola** (Po), so-la
**solbær** (Nw), sol-bær
**sole** (US), sole
**sole à la bonne femme** (Fr),
sole à la bonne femme
**sole à la dieppoise** (Fr), sole
à la diep-poise

*sole à la Dugléré* (Fr), sole à la Dug-lé-ré

*sole à la normande* (Fr), sole à la nor-mande

*sole à la meunière* (Fr), sole à la meu-niè-re

*sole cardinal* (Fr), sole car-di-nal

*sole colbert* (Fr), sole col-bert

*sole Marguery* (Fr), sole Mar-gue-ry

*soles de douvres* (Fr), soles de dou-vres

*sólet* (Hu), só-let

*solianka* (Rs), sol-ian-ka

*solid fat index* (US), sol-id fat in-dex

*sologa* (Sw), so-lo-ga

*solomillo* (Sp), so-lo-mi-llo

*somen* (Jp), so-men

*somlói galuskas* (Hu), som-lói ga-lus-kas

*sondesh* (Ia), son-desh

*sonhos* (Pg), so-nhos

*sonka* (Hu), son-ka

*sonkás metélt* (Hu), son-kás me-télt

*sonkás palacsinták* (Hu), son-kás pa-la-csin-ták

*sonkás rétes* (Hu), son-kás ré-tes

*sood manti* (Tr), sood man-ti

*soon* (In), soon

*soong* (Tr), soong

*soongi aghtsan* (Tr), soongi agh-tsan

*soop* (Rs), soop

*soorj* (Tr), soorj

*sop* (In), sop

*sopa* (Sp, Pg), so-pa

*sopa à alentejana* (Pg), so-pa à a-len-te-ja-na

*sopa al jerez* (Sp), so-pa al je-rez

*sopa al queso* (Sp), so-pa al que-so

*sopa borracha* (Sp), so-pa bo-rra-cha

*sopa de albóndigas* (Sp), so-pa de al-bón-di-gas

*sopa de ameijoas* (Pg), so-pa de a-me-i-jo-as

*sopa de batata e agrião* (Pg), so-pa de ba-ta-ta e a-gri-ão

*sopa de camarão* (Pg), so-pa de ca-ma-rão

*sopa de fideos* (Sp), so-pa de fi-de-os

*sopa de feijão* (Pg), so-pa de fe-i-jã-o

*sopa de frijol negro* (Sp), so-pa de fri-jol ne-gro

*sopa de grão* (Pg), so-pa de grã-o

*sopa de hortaliça* (Pg), so-pa de hor-ta-li-ça

*sopa de legumbres con huevos* (Sp), so-pa de le-gum-bres con hue-vos

*sopa de mariscos* (Pg, Sp), so-pa de ma-ris-cos

*sopa de mexilhão* (Pg), so-pa de me-xi-lhão

*sopa de tomate à alentejana* (Pg), so-pa de to-ma-te à a-len-te-ja-na

*sopaipillas* (Sp), so-pai-pi-llas

*sopaipillas chilenitas* (Sp), so-pai-pi-llas chi-le-ni-tas

*sopa seca* (Sp), so-pa se-ca

*sopa seca de fideos* (Sp), so-pa se-ca de fi-de-os

*sopa transmontana* (Pg), so-pa trans-mon-ta-na

*sopes* (Sp), so-pes

*sopita* (Sp), so-pi-ta

*sopón de garbanzos con patas de cerdo* (Sp), so-pón de gar-ban-zos con pa-tas de cer-do

*sopp* (Nw), sopp

*soppstuing* (Nw), sopp-stu-ing

*sör* (Hu), sör

*soramame* (Jp), so-ra-ma-me

*sorbet* (Fr), sor-bet

*sorbet au cassis* (Fr), sor-bet au cas-sis

*Sorbett* (Gr), Sor-bett

*sorbetto* (It), sor-bet-to

*sorbitol* (US), sor-bi-tol

*sorghum flour* (US), sor-ghum flour

*sorghum syrup* (US), sor-ghum syr-up

*sorrel* (US), sor-rel

*sorvete* (Pg), sor-ve-te

*sos* (Po, Tr), sos

*sosaties* (Af), so-sat-ies

*sosej* (Af-Swahili), so-sej

*sōsēji* (Jp), sō-sē-ji

*sosis* (Tr), so-sis

*Sosse* (Gr), Sos-se

*søsterkage* (Da), søs-ter-ka-ge

*sōsu* (Jp), sō-su

*sos z jaj do ryby* (Po), sos z jaj do ry-by

*sotirano* (SC), so-ti-ra-no

*soto* (In), so-to

*soto ayam* (In), so-to a-yam

*sött* (Sw), sött

*sottaceti* (It), sot-ta-ce-ti

*søtunge* (Da), sø-tun-ge

*soubise* (Fr), sou-bise

*soubise, sauce* (Fr), sou-bise, sauce

*sou boereg* (Tr), sou boe-reg

*souci* (Fr), sou-ci

*soufflé* (Fr), souf-flé

*soufflé ambassadrice* (Fr), souf-flé am-bas-sa-drice

*soufflé aux épinards* (Fr), souf-flé aux é-pi-nards

*soufflé aux fraises* (Fr), souf-flé aux frai-ses

*soufflé ze śledzia* (Po), souf-flé ze śle-dzia

*soup* (US), soup

*soupa avgolemono* (Gk), sou-pa av-go-le-mo-no

*soupe* (Fr), soupe

*soupe à la reine* (Fr), soupe à la reine

*soupe à l'oignon* (Fr), soupe à l'oi-gnon

*soupe au pistou* (Fr), soupe au pis-tou

*soupe aux congres* (Fr), soupe aux con-gres

*soupe aux marrons* (Fr), soupe aux mar-rons

*soupe bonne femme* (Fr), soupe bonne femme

*soupe de jour* (Fr), soupe de jour

*sour cherry* (US), sour cher-
ry
*sour cream* (US), sour cream
*sourdough* (US), sour-dough
*sour orange* (US), sour or-
ange
*soursop* (US), sour-sop
*souse* (US), souse
*soused* (GB), soused
*sous la cendre* (Fr), sous la
cen-dre
*Southern fried chicken* (US),
South-ern fried chick-en
*souvlakia* (Gk), souv-la-kia
*soybean* (US), soy-bean
*soybean curd* (US), soy-bean
curd
*soy sauce* (US), soy sauce
*spagety* (Cz), spa-ge-ty
*spaghetti* (It), spa-ghet-ti
*spaghettini* (It), spa-ghet-ti-ni
*spalla di vitello al forno* (It),
spal-la di vi-tel-lo al for-no
*spanać* (SC), spa-nać
*spanakopitta* (Gk), spa-na-
ko-pit-ta
*spanakorizo* (Gk), spa-na-
ko-ri-zo
*španělští ptáčci* (Cz), špa-
něl-ští ptáč-ci
*Spanferkel* (Gr), Span-fer-kel
*spanischer Pfeffer* (Gr), spa-
ni-scher Pfef-fer
*sparagio* (It), spa-ra-gio
*spárga* (Hu), spár-ga
*spárgaleves* (Hu), spár-ga-le-
ves
*Spargel* (Gr), Spar-gel
*Spargelkohl* (Gr), Spar-gel-
kohl

*špargle* (SC), špar-gle
*sparris* (Sw), spa-rris
*spatula* (US), spat-u-la
*Spätzle* (Gr), Spätz-le
*Speck mit Eiern* (Gr), Speck
mit Ei-ern
*speculaas* (Du), spe-cu-laas
*spegesild* (Da), spe-ge-sild
*spegepølse* (Da), spe-ge-
pøl-se
*speilegg* (Nw), speil-egg
*spek* (Du), spek
*spekekjøtt* (Nw), spe-ke-
kjøtt
*speldings* (Sc), speld-ings
*spenat* (Sw), spe-nat
*spenót* (Hu), spe-nót
*spersiebonen* (Du), sper-sie-
bo-nen
*spetsiotiko* (Gk), spe-tsio-ti-
ko
*spettakaka* (Sw), spet-ta-ka-
ka
*spezzatino di vitello* (It),
spez-za-ti-no di vi-tel-lo
*spice* (US), spice
*spickgans* (Gr), spick-gans
*spiedini* (It), spie-di-ni
*Spiegeleier* (Gr), Spie-gel-ei-
er
*Spieseeis* (Gr), Spie-se-eis
*spiess* (Gr), spiess
*spigola* (It), spi-go-la
*spinach* (US), spin-ach
*spinach beet* (US), spin-ach
beet
*spinaci* (It), spi-na-ci
*spinaci alla romana* (It),
spi-na-ci al-la ro-ma-na
*spinat* (Nw), spi-nat

*Spinat* (Gr), Spi-nat

*spinazie* (Du), spi-na-zie

*spiny lobster* (US), spin-y lob-ster

*spitisies hilopites* (Gk), spi-ti-si-es hi-lo-pi-tes

*sponge cake* (US), sponge cake

*spoon bread* (US), spoon bread

*sporacciona, alla* (It), spo-rac-ci-o-na, al-la

*spotted dog* (GB), spot-ted dog

*sprat* (US), sprat

*Springerle* (Gr), Sprin-ger-le

*spring roll* (US), spring roll

*spritsar* (Sw), sprit-sar

*Spritzgeback* (Gr), Spritz-ge-back

*Spritzkuchen* (Gr), Spritz-ku-chen

*sprot* (Du), sprot

*Sprotte* (Gr), Sprot-te

*sproty* (Cz), spro-ty

*spruce beer* (US), spruce beer

*spruitjes* (Du), sprui-tjes

*spud* (US), spud

*spuitwater* (Du), spuit-wa-ter

*spuma* (It), spu-ma

*spuma di banane* (It), spu-ma di ba-na-ne

*spumone* (It), spu-mo-ne

*squab* (US), squab

*square cut shoulder* (US), square cut should-er

*squid* (US), squid

*squirrel* (US), squir-rel

*srce* (SC), sr-ce

*srdce* (Cz), srd-ce

*średnio* (Po), śred-nio

*srnčí maso* (Cz), srn-čí ma-so

*srpski sir* (SC), srp-ski sir

*stabilizers* (US), sta-bi-liz-ers

*stafilia* (Gk), sta-fi-lia

*stalk* (US), stalk

*standing rump* (US), stand-ing rump

*stap* (Sc), stap

*star anise* (US), star an-ise

*star fruit* (US), star fruit

*stark senap* (Sw), stark se-nap

*starkt kryddat* (Sw), starkt kry-ddat

*šťáva z masa* (Cz), šťá-va z ma-sa

*steak au poivre* (Fr), steak au poi-vre

*steak tartare* (Fr), steak tar-tare

*steam, to* (US), steam, to

*stebghin* (Tr), steb-ghin

*stebghini aghtsan* (Tr), steb-ghini agh-tsan

*Steckrube* (Gr), Steck-ru-be

*steep* (US), steep

*stefado* (Gk), ste-fa-do

*stein wines* (Gr), stein wines

*stek* (Nw), stek

*stekt* (Sw), stekt

*stekt på spett* (Sw), stekt på spett

*stekt sill eller stromming* (Sw), stekt sill el-ler strom-ming

*stellini* (It), stel-li-ni

*steur* (Du), steur

*stew, to* (US), stew, to

*stiacciata* (It), sti-ac-cia-ta

*štika* (Cz), šti-ka

*stikkelsbær* (Nw), stikk-els-bær

*stikkelsbærgrød* (Da), stik-kels-bær-grød

*Stilton* (GB), Stil-ton

*stingaree* (US), sting-a-ree

*Stinte mit saurer Sosse* (Gr), Stin-te mit sau-rer Sos-se

*stir, to* (US), stir, to

*stirred custard* (US), stirred cus-tard

*stoccafisso* (It), stoc-ca-fis-so

*stoccafisso accomodato* (It), stoc-ca-fis-so ac-co-mo-da-to

*stock* (US), stock

*stockfish* (US), stock-fish

*stockpot* (US), stock-pot

*Stollen* (Gr), Stol-len

*stör* (Sw), stör

*stør* (Da, Nw), stør

*storione* (It), sto-ri-o-ne

*storskate* (Nw), stor-ska-te

*stracchino* (It), strac-chi-no

*stracciatella* (It), strac-cia-tel-la

*stracotto* (It), stra-cot-to

*strain, to* (US), strain, to

*Straussburg sausage* (US), Strauss-burg sau-sage

*straw mushrooms* (US), straw mush-rooms

*Streuselkuchen* (Gr), Streu-sel-ku-chen

*striped bass* (US), striped bass

*strip steak* (US), strip steak

*strömming* (Sw), ström-ming

*štruca od mesa* (SC), štru-ca od me-sa

*Stückchen* (Gr), Stück-chen

*stufato* (It), stu-fa-to

*stufato di manzo* (It), stu-fa-to di man-zo

*sturgeon* (US), stur-geon

*sturgeon roe* (US), stur-geon roe

*su* (Tr), su

*su* (Jp), su

*suar kaa maans* (Ia), suar kaa maans

*sucadekoek* (Du), su-ca-de-koek

*succo* (It), suc-co

*succo d'arancio* (It), suc-co d'a-ran-cio

*succo di frutta* (It), suc-co di frut-ta

*succotash* (US), suc-co-tash

*sucées* (Fr), su-cées

*suchar* (Cz), su-char

*suco* (Pg), su-co

*suco de laranja* (Pg), su-co de la-ran-ja

*sucre* (Fr), sucre

*sudak* (Rs), su-dak

*sudako* (Jp), su-da-ko

*suero de la leche* (Sp), su-e-ro de la le-che

*suet* (GB), su-et

*suffle* (It), suf-fle

*sufle* (Tr), su-fle

*sugar* (GB), sug-ar

*sugar-free* (US), sug-ar-free

*sugar peas* (US), sug-ar peas

*sugo* (It), su-go

*suika* (Jp), su-i-ka

*suiker* (Du), sui-ker

*suimitsu* (Jp), su-i-mit-su

*suimono* (Jp), su-i-mo-no

*suji malpua* (Ia), su-ji mal-pua

*sukari* (Af-Swahili), su-ka-ri

*sukhdor* (Tr), sukh-dor

*sukhdori tahtsan* (Tr), sukh-dori tah-tsan

*sukhdori yev* (Tr), sukh-dori yev

*suki* (Jp), su-ki

*sukiyaki* (Jp), su-ki-ya-ki

*sukkar* (Ar), suk-kar

*sukker* (Nw), suk-ker

*sůl* (Cz), sůl

*süllö* (Hu), sül-lö

*sultana* (US), sul-tan-a

*sülün* (Tr), sü-lün

*sumashijiru* (Jp), su-ma-shi-ji-ru

*summer pudding* (GB), sum-mer pud-ding

*summer sausage* (US), sum-mer sau-sage

*sumomo* (Jp), su-mo-mo

*sumpoog* (Tr), sum-poog

*sumpoogi aboor* (Tr), sum-poogi a-boor

*sumpoogi leetsk* (Tr), sum-poogi leetsk

*sumpoogi tahtsan* (Tr), sum-poogi tah-tsan

*sumpoogi yev loligi aghtsan* (Tr), sum-poogi yev lo-ligi agh-tsan

*sumpoogov yev meesov pancharegan* (Tr), sum-poog-ov yev mees-ov pan-cha-re-gan

*sunchoke* (US), sun-choke

*sundae* (US), sun-dae

*Sunderland pudding* (US), Sun-der-land pud-ding

*sun drying* (US), sun dry-ing

*sunflower oil* (US), sun-flow-er oil

*šunka* (Cz, SC), šun-ka

*sunomono* (Jp), su-no-mo-no

*suola* (Fi), su-o-la

*suolasilakka* (Fi), su-o-la-si-lak-ka

*suomuurain* (Fi), su-o-muu-rain

*supa avgholemono* (Gk), su-pa av-gho-le-mo-no

*suppe* (Nw), sup-pe

*Suppe* (Gr), Sup-pe

*suppli* (It), sup-pli

*suppli di riso* (It), sup-pli di ri-so

*suppon* (Jp), sup-pon

*supreme sauce* (Fr), su-preme sauce

*suprêmes de volaille à blanc* (Fr), su-prêmes de vol-aille à blanc

*supu ya kuku* (Af-Swahili), su-pu ya ku-ku

*supu ya mboga* (Af-Swahili), su-pu ya m-bo-ga

*suquet de peix* (Sp), su-quet de pe-ix

*sur commande* (Fr), sur com-mande

*sur grädde* (Sw), sur gräd-de

*surimi* (Jp), su-ri-mi

*surkål* (Nw, Sw), sur-kål

*sur melk* (Nw), sur melk

*surströmming* (Sw), sur-ström-ming

**sušené švestky** (Cz), su-še-né švest-ky
**sušenky** (Cz), su-šen-ky
**sushi** (Jp), su-shi
**sushi-meshi** (Jp), su-shi-me-shi
**Süsstoff** (Gr), Süs-stoff
**süt** (Tr), süt
**sutēki** (Jp), su-tē-ki
**sütve** (Hu), süt-ve
**suutarinilohi** (Fi), suu-ta-rin-i-lo-hi
**suve šljive** (SC), su-ve šlji-ve
**suvlas** (Gk), suv-las
**suzuke** (Jp), su-zu-ke
**suzuki** (Jp), su-zu-ki
**suzuki shioyaki** (Jp), su-zu-ki shi-o-ya-ki
**svamp** (Sw), svamp
**svarta vinbär** (Sw), svar-ta vin-bär
**svecia ost** (Sw), sve-ci-a ost
**sveitserost** (Nw), sveit-ser-ost
**švestky** (Cz), švest-ky
**svíčková na smetaně** (Cz), svíč-ko-vá na sme-ta-ně
**svinekjøtt** (Nw), svi-ne-kjøtt
**svinemørbrad** (Da), svi-ne-mør-brad
**svinjetina** (SC), svi-nje-ti-na
**svisker** (Nw), svisk-er
**svoja** (SC), svo-ja
**swàn** (Ch), swàn
**swān-syǎu-tsài** (Ch), swān-syǎu-tsài
**sweat** (GB), sweat
**swede** (GB), swede
**sweet-and-sour** (US), sweet-and-sour
**sweetbread** (US), sweet-bread

**sweet chocolate** (US), sweet cho-co-late
**sweet cicely** (GB), sweet cic-e-ly
**sweet corn** (US), sweet corn
**sweet pepper** (US), sweet pep-per
**sweet potato** (US), sweet po-ta-to
**sweet rice flour** (US), sweet rice flour
**sweet sop** (GB), sweet sop
**Swiss chard** (US), Swiss chard
**Swiss cheese** (US), Swiss cheese
**Swiss roll** (GB), Swiss roll
**Swiss sausage** (US), Swiss sau-sage
**Swiss steak** (US), Swiss steak
**swizzle** (GB), swiz-zle
**swordfish** (US), sword-fish
**syán-dàn** (Ch), syán-dàn
**syāng-jyǎu** (Ch), syāng-jyǎu
**syāng-tsài** (Ch), syāng-tsài
**syāng-yóu** (Ch), syāng-yóu
**syǎu-mài** (Ch), syǎu-mài
**syǎu-nyóu-ròu** (Ch), syǎu-nyóu-ròu
**syǎu-syǎr** (Ch), syǎu-syǎr
**syǎu-yáng-ròu** (Ch), syǎu-yáng-ròu
**sybo** (Sc), sy-bo
**sydän** (Fi), sy-dän
**syī-gwā** (Ch), syī-gwā
**syìng-dz** (Ch), syìng-dz
**syìng-rén** (Ch), syìng-rén
**syìng-rén dòu-fú** (Ch), syìng-rén dòu-fú

**syllabub** (GB), syl-la-bub
**sylt** (Sw), sylt
**sylte** (Da), syl-te
**syltesild** (Da), syl-te-sild
**syltetøy** (Nw), syl-te-tøy
**sýr** (Cz), sýr
**syrniki** (Rs), syr-ni-ki
**syrup** (US), syr-up
**sywě-dòu** (Ch), sywě-dòu
**szalonnás gombóc** (Hu), sza-lon-nás gom-bóc
**szalonnás súlt** (Hu), sza-lon-nás súlt
**szardinia** (Hu), szar-di-nia
**szarvas** (Hu), szar-vas
**Sz-chwān hú-jyāu** (Ch), Sz-chwān hú-jyāu
**Sz-chwān jyău-dz** (Ch), Sz-chwān jyău-dz

**Szechuan cooking** (Ch), Sze-chu-an cook-ing
**Szechuan pepper** (Ch), Sze-chu-an pep-per
**székelygulyás** (Hu), szé-ke-ly-gu-lyás
**székely sertésborda** (Hu), szé-ke-ly ser-tés-bor-da
**szeletek** (Hu), sze-le-tek
**szilva** (Hu), szil-va
**szilvalekvár** (Hu), szil-va-lek-vár
**szilvás gombóc** (Hu), szil-vás gom-bóc
**sziv** (Hu), sziv
**szölö** (Hu), szö-lö
**szparagi** (Po), szpa-ra-gi
**szpinak** (Po), szpi-nak
**szprot** (Po), szprot
**sztufada** (Po), sztu-fa-da

*T*

**taartjes** (Du), taar-tjes
**taateleita** (Fi), taa-te-lei-ta
**taazaa** (Ia), taa-zaa
**tabasco** (Sp), ta-bas-co
**Tabasco sauce** (US), Ta-bas-co sauce
**tabbouleh** (Ar), tab-bou-leh
**tabboun** (Ar), tab-boun
**tabeet** (Jw), ta-beet

**table cream** (US), ta-ble cream
**table d'hote** (Fr), tab-le d'hote
**table salt** (US), ta-ble salt
**table water** (US), ta-ble wa-ter
**table wine** (US), ta-ble wine
**tablier de sapeur** (Fr), tab-li-er de sap-eur

*Tâche* (Fr), Tâche
*tacchino* (It), tac-chi-no
*tacchino ripieno* (It), tac-chi-no ri-pie-no
*Táchira* (Sp), Tá-chi-ra
*tack* (GB), tack
*taco* (Mx), ta-co
*tacon* (Fr), ta-con
*tacos de pescado* (Mx), ta-cos de pes-ca-do
*tacos de pollo asado* (Mx), ta-cos de po-llo a-sa-do
*tädinkakut* (Fi), tä-din-ka-kut
*tädin kalavuoka* (Fi), tä-din ka-la-vu-o-ka
*Tafelbirne* (Gr), Ta-fel-bir-ne
*Tafelspitz* (Gr), Ta-fel-spitz
*Tafelwein* (Gr), Ta-fel-wein
*Taffel* (Da), Taf-fel
*Taffelost* (Nw), Taf-fel-ost
*taffy* (US), taf-fy
*taffy* (GB), taf-fy
*Tafi* (Sp), Ta-fi
*Tagessuppe* (Gr), Ta-ges-sup-pe
*tagine* (Ar), ta-gine
*tagin orz* (Ar), ta-gin orz
*tagliarini* (It), ta-glia-ri-ni
*tagliatelle* (It), ta-glia-tel-le
*tagliatelle al ragù* (It), ta-glia-tel-le al ra-gù
*tagliatelle verde* (It), ta-glia-tel-le ver-de
*tahari* (Ia), ta-ha-ri
*taheenov dzaghgagaghamp* (Tr), ta-heen-ov dzagh-ga-ga-ghamp
*taheenov gargantag* (Tr), ta-heen-ov gar-gan-tag

*taheenov jajegh* (Tr), ta-heen-ov ja-jegh
*taheen yev rube* (Tr), ta-heen yev rube
*tahina* (Ar), ta-hi-na
*tahini* (Ar), ta-hi-ni
*tahiyn* (Ar), ta-hiyn
*tahn* (Tr), tahn
*tahtsan* (Tr), tah-tsan
*tahu* (In), tahu
*tahu campur* (In), tahu cam-pur
*tai* (Jp), tai
*taiglach* (Jw), taig-lach
*tailor* (US), tai-lor
*tails* (US), tails
*taimen* (Fi), ta-i-men
*taimeshi* (Jp), tai-me-shi
*tai nam* (Vt), tai nam
*tai no sashimi* (Jp), tai no sa-shi-mi
*tai tempura* (Jp), tai tem-pu-ra
*tajine* (Ar), ta-jine
*tajine ez zitoun* (Ar), ta-jine ez zi-toun
*takenoko* (Jp), ta-ke-no-ko
*takenokomeshi* (Jp), ta-ke-no-ko-me-shi
*take out* (US), take out
*taki* (Jp), ta-kī
*takikomi-gohan* (Jp), ta-ki-ko-mi-go-han
*taklia* (Ar), takl-ia
*tako* (Jp), ta-ko
*tako kushisashi* (Jp), ta-ko ku-shi-sa-shi
*takrai* (Th), tak-rai
*takuan* (Jp), ta-ku-an

*takuan maki* (Jp), ta-ku-an ma-ki

*takuan zuke* (Jp), ta-ku-an zu-ke

*tala bua* (Ia), ta-la hua

*talas* (In), ta-las

*Taleggio* (It), Ta-leg-gio

*tali machchi* (Ia), ta-li mach-chi

*tallarines* (Sp), ta-lla-ri-nes

*Tallyrand* (Fr), Tal-ly-rand

*talmouse* (Fr), tal-mouse

*tamaatim* (Ar), ta-maa-tim

*tamago* (Jp), ta-ma-go

*tamagodōfu* (Jp), ta-ma-go-dō-fu

*tamagodon* (Jp), ta-ma-go-don

*tamago toji udon* (Jp), ta-ma-go to-ji u-don

*tamagoyaki* (Jp), ta-ma-go-ya-ki

*tamal de cazuela* (MX), ta-mal de ca-zu-e-la

*tamale* (Mx), ta-ma-le

*tamale pie* (US), ta-ma-le pie

*tamalitos* (Mx), ta-ma-li-tos

*tamalitos de elote* (Mx), ta-ma-li-tos de e-lo-te

*tamanegi* (Jp), ta-ma-ne-gi

*tâmaras* (Pg), tâ-ma-ras

*tamari* (Jp), ta-ma-ri

*tamarind* (Ia), ta-ma-rind

*tamattar* (Ia), ta-mat-tar

*tamattar chatni* (Ia), ta-mat-tar chat-ni

*Tambo* (Aa), Tam-bo

*Tamié* (Fr), Ta-mié

*tamis* (Fr), tam-is

*tammy* (US), tam-my

*tampala* (Ia), tam-pa-la

*Tandil* (Sp), Tan-dil

*tandir kebab* (Tr), tan-dir ke-bab

*tandoor* (Ia), tan-door

*tandoori* (Ia), tan-doo-ri

*tandoori murg* (Ia), tan-doo-ri murg

*tandoori roti* (Ia), tan-doo-ri ro-ti

*táng* (Ch), táng

*tāng* (Ch), tāng

*tángcù liji* (Ch), tángcù liji

*táng-dz* (Ch), táng-dz

*tangelo* (US), tan-ge-lo

*tangerine* (US), tan-ge-rine

*tang kwah ah jad* (Th), tang kwah ah jad

*tangmiàn* (Ch), tang-miàn

*tango* (Af-Swahili), ta-ngo

*tangor* (US), tan-gor

*tanmen* (Jp), tan-men

*tanner* (US), tan-ner

*tansansui* (Jp), tan-san-sui

*tanuki jiru* (Jp), ta-nu-ki ji-ru

*tanuki soba* (Jp), ta-nu-ki so-ba

*tanuki udon* (Jp), ta-nu-ki u-don

*Tanzenberger* (Gr), Tan-zen-ber-ger

*taoge* (In), tao-ge

*Tapachula* (Sp), Ta-pa-chu-la

*tapada de pollo* (Mx), ta-pa-da de po-llo

*tapai* (In), ta-pai

*tapang baka* (Ph), ta-pang ba-ka

*tapang baboy* (Ph), ta-pang ba-boy

*tapas* (Sp), ta-pas

*tapenade* (Fr), ta-pe-nade

*tapinette* (Fr), ta-pi-nette

*tapioca* (US), tap-i-o-ca

*täpläsilli* (Fi), tä-plä-sil-li

*taquitos* (Mx), ta-qui-tos

*tara* (Jp), ta-ra

*Tara* (GB), Ta-ra

*tarako* (Jp), ta-ra-ko

*taramasalata* (Gk), ta-ra-ma-sa-la-ta

*taratoor* (Tr), tar-a-toor

*tarator* (Bl), tar-a-tor

*taratur* (Ar), tar-a-tur

*tarbooz* (Ia), tar-booz

*tarhana çorbasi* (Tr), tar-ha-na çor-ba-si

*tarhonya* (Hu), tar-ho-nya

*tari* (Ia), ta-ri

*tarka* (Ia), tar-ka

*tarkari* (Ia), tar-ka-ri

*taro* (US), ta-ro

*tarragon* (US), tar-ra-gon

*tarragon butter* (Fr), tar-ra-gon but-ter

*tart* (US), tart

*tarta* (Sp), tar-ta

*tårta* (Sw), tår-ta

*tartare* (Fr), tar-tare

*Tartare* (Fr), Tar-tare

*tartar med aeg* (Da), tar-tar med aeg

*tartar sauce* (US), tar-tar sauce

*tartar steak* (US), tar-tar steak

*tarte* (Fr), tarte

*tarte à l'oignon* (Fr), tarte à l'oi-gnon

*tarte alsacienne* (Fr), tarte al-sa-cienne

*tarte aux fruits* (Fr), tarte aux fruits

*tarte des demoiselles Tatin* (Fr), tarte des dem-oi-sel-les Ta-tin

*tartelette* (Fr), tarte-lette

*tartelette Agnes* (Fr), tarte-lette A-gnes

*tartelette à la Florentine* (Fr), tarte-lette à la Flo-ren-tine

*tartelette au Eglefin* (Fr), tarte-lette au E-gle-fin

*tarteletter med hummer og asparges* (Da), tar-te-let-ter med hum-mer og a-sparges

*tarte liègeoise* (Fr), tarte liè-ge-oise

*tartina* (It), tar-ti-na

*tartines* (Fr), tar-tines

*tartufi* (It), tar-tu-fi

*tarwe brood* (Du), tar-we brood

*Tascherin* (Gr), Ta-sche-rin

*tas kebob* (Tr), tas ke-bob

*tassergal* (Fr), tas-ser-gal

*tataki gobō* (Jp), ta-ta-ki go-bō

*tatli* (Tr), tat-li

*Tätschii* (Gr), Tät-schii

*tatti* (Fi), tat-ti

*tatties* (Sc), tat-ties

*Taube* (Gr), Tau-be

*Tauben in Specksauce* (Gr), Tau-ben in Speck-sau-ce

*taucheo* (Ml), tau-cheo

*taucho* (In, Th), tau-cho

*táu-dz* (Ch), táu-dz
*tauşan* (Tr), tau-şan
*tausi* (Ph), tau-si
*tautog* (US), tau-tog
*tavada* (Tr), ta-va-da
*tavola fredda* (It), ta-vo-la fred-da
*tavuk* (Tr), ta-vuk
*tavuk çorbasi* (Tr), ta-vuk çor-ba-si
*tavuk izgara* (Tr), ta-vuk iz-ga-ra
*tavuk suyu* (Tr), ta-vuk su-yu
*tay* (Tr), tay
*täytekakku* (Fi), tä-y-te-kak-ku
*tă-yú* (Ch), tă-yú
*taze sebze* (Tr), ta-ze seb-ze
*T-bone steak* (US), T-bone steak
*te* (Da, Nw, Sw), te
*tè* (It), tè
*té* (Sp), té
*tea* (US), tea
*teaberry* (US), tea-ber-ry
*teal* (US), teal
*tebrød* (Nw), te-brød
*tee* (Fi), tee
*Tee* (Gr), Tee
*teen* (Ar), teen
*teeri* (Fi), tee-ri
*teesri* (Ia), tee-sri
*teetar* (Ia), tee-tar
*teff* (Ia), teff
*tefteli* (Rs), tef-te-li
*teh* (In), teh
*Téiggemüse* (Gr), Téig-ge-mü-se
*tej* (Hu), tej

*tejfeles úburgonya* (Hu), tej-fe-les ú-bur-go-nya
*tej patta* (Ia), tej pat-ta
*tejszines* (Hu), tej-szi-nes
*tekaka* (Sw), te-ka-ka
*tekka miso* (Jp), tek-ka mi-so
*tél* (Ia), tél
*tel beneer* (Tr), tel be-neer
*teleci* (Cz), te-le-ci
*teleci kotleta* (Cz), te-le-ci kot-le-ta
*Teleme* (Gk), Te-lem-e
*Teleme Jack* (US), Te-lem-e Jack
*teletina* (SC), te-le-ti-na
*tel khadayeef* (Tr), tel kha-da-yeef
*tellin* (US), tel-lin
*telur* (In), te-lur
*telur goreng* (In), te-lur go-reng
*telur rebus* (In), te-lur re-bus
*telyatina* (Rs), tel-ya-ti-na
*temesvári sertésborda* (Hu), te-mes-vá-ri ser-tés-bor-da
*tempe* (In), tem-pe
*temper* (US), tem-per
*Temple orange* (US), Tem-ple or-ange
*tempura* (Jp), tem-pu-ra
*tempura soba* (Jp), tem-pu-ra so-ba
*tench* (US), tench
*tende* (Af-Swahili), te-nde
*tenderette* (US), ten-der-ette
*tenderloin* (US), ten-der-loin
*tendon* (Jp), ten-don
*tendron de veau* (Fr), ten-dron de veau

**tengeri rák** (Hu), ten-ge-ri rák

**tentsuyu** (Jp), ten-tsu-yu

**tepertös pogácsa** (Hu), te-per-tös po-gá-csa

**tepid** (US), tep-id

**teppenyaki** (Jp), tep-pen-ya-ki

**tequilla** (Mx), te-qui-lla

**tereyaği** (Tr), te-re-ya-ği

**teri** (Jp), te-ri

**terik ayam** (In), terik a-yam

**terik tempe** (In), terik tem-pe

**teriyaki** (Jp), te-ri-ya-ki

**ternera** (Sp), ter-ne-ra

**ternera al jerez** (Sp), ter-ne-ra al je-rez

**ternera borracha** (Sp), ter-ne-ra bo-rra-cha

**ternera en agujas** (Sp), ter-ne-ra en a-gu-jas

**ternera jardinera** (Sp), ter-ne-ra jar-di-ne-ra

**terrapin** (US), ter-ra-pin

**terrine** (Fr), ter-rine

**terrine de caneton** (Fr), ter-rine de ca-ne-ton

**terrine maison** (Fr), ter-rine mai-son

**tertanoosh** (Tr), tert-a-noosh

**Tête de Moine** (Fr), Tête de Moine

**tête de veau** (Fr), tête de veau

**tetrazzini** (It), te-traz-zi-ni

**Teufelsdreck** (Gr), Teu-fels-dreck

**textured plant protein** (US), tex-tured plant pro-tein

**thali** (Ia), tha-li

**thalj** (Ar), thalj

**thandai** (Ia), than-dai

**thé** (Fr), thé

**thé à la menthe** (Fr), thé à la menthe

**thee** (Du), thee

**thee complet** (Du), thee com-plet

**theobromine** (US), the-o-bro-mine

**theophylline** (US), the-o-phyl-line

**thermidor** (Fr), therm-i-dor

**thiamine** (US), thi-a-mine

**thicken** (US), thick-en

**thickener** (US), thick-en-er

**thimbleberry** (US), thim-ble-ber-ry

**thit** (Vt), thit

**Thompson seedless** (US), Thomp-son seed-less

**thon** (Fr), thon

**Thousand Island dressing** (US), Thou-sand Is-land dress-ing

**thousand-year eggs** (US), thou-sand-year eggs

**thread** (US), thread

**Thunfisch** (Gr), Thun-fisch

**Thunfischsalat** (Gr), Thun-fisch-sa-lat

**Thuringer** (Gr), Thu-rin-ger

**thym** (Fr), thym

**thyme** (US), thyme

**Tia Maria** (Sp), Ti-a Ma-ri-a

**tibid** (Da), ti-bid

**tien mien jiàng** (Ch), tien mien jiàng

**tif-fah** (Ar), tif-fah

**tiger lily buds** (US), ti-ger lil-y buds

*tiges* (Fr), tiges
*tighanites* (Gk), ti-gha-ni-tes
*Tignard* (Fr), Ti-gnard
*tikki* (Ia), tik-ki
*tikki chana dal* (Ia), tik-ki cha-na dal
*tikvice* (SC), tik-vi-ce
*til* (Ia), til
*ti leaves* (US), ti leaves
*tilefish* (US), tile-fish
*Tillamook* (US), Till-a-mook
*tilli* (Fi), til-li
*tillikastike* (Fi), til-li-kas-ti-ke
*tilliliha* (Fi), til-li-li-ha
*tillisilakka* (Fi), til-li-si-lak-ka
*tilltug* (Sw), till-tug
*Tilsit* (Gr), Til-sit
*timbale* (Fr), tim-bale
*timballo abruzzi* (It), tim-bal-lo a-bruz-zi
*timballo di riso con salsicce* (It), tim-bal-lo di ri-so con sal-sic-ce
*tim joke* (Ch), tim joke
*timo* (It), ti-mo
*tinda* (Ia), tin-da
*tini* (Af-Swahili), ti-ni
*Tintenfisch* (Gr), Tin-ten-fisch
*tippaleivät* (Fi), tip-pa-lei-vät
*tipsy cake* (GB), tip-sy cake
*tiram* (In), ti-ram
*tiramisu* (It), ti-ra-mi-su
*tiri* (Gk), ti-ri
*Tirolen Eierspeisen* (Gr), Ti-ro-len Ei-er-spei-sen
*Tiroler Knödeln* (Gr), Ti-ro-ler Knö-deln
*tiropeta* (Gk), ti-ro-pe-ta

*tirotrigona* (Gk), ti-ro-tri-go-na
*tisane* (Fr), ti-sane
*titori bhujia* (Ia), ti-to-ri bhu-jia
*tlačenka* (Cz), tla-čen-ka
*tlami* (Ar), tla-mi
*tmar* (Ar), tmar
*tmar michi* (Ar), tmar mi-chi
*toad-in-the-hole* (GB), toad-in-the-hole
*toast* (US), toast
*tocino* (Sp), to-ci-no
*tocino de cielo* (Sp), to-ci-no de ci-e-lo
*tocopherols* (US), to-coph-er-ols
*toddy* (GB), tod-dy
*tofu* (Jp), to-fu
*togan* (Jp), to-gan
*togarashi* (Jp), to-ga-ra-shi
*tohm* (Ar), tohm
*tojás* (Hu), to-jás
*tojáskocsonya* (Hu), to-jás-ko-cso-nya
*tokány* (Hu), to-ká-ny
*Tokay grape* (US), To-kay grape
*tökebal* (Hu), tö-ke-hal
*tökfözelék* (Hu), tök-fö-ze-lék
*Toll House cookie* (US), Toll House cook-ie
*töltött* (Hu), töl-tött
*töltött káposzta* (Hu), töl-tött ká-posz-ta
*töltött paprika* (Hu), töl-tött pap-ri-ka
*töltött paradicsom* (Hu), töl-tött pa-ra-di-csom
*tomaatti* (Fi), to-maat-ti

**tomaattimehu** (Fi), to-maat-ti-me-hu

**tomaattisilakka** (Fi), to-maat-ti-si-lak-ka

**tomalley** (US), tom-al-ley

**Tom and Jerry** (GB), Tom and Jer-ry

**tomat** (Da), to-mat

**tomate** (Pg, Sp), to-ma-te

**Tomaten** (Gr), To-ma-ten

**Tomatensalat** (Gr), To-ma-ten-sa-lat

**tomatensap** (Du), to-ma-ten-sap

**Tomatensuppe** (Gr), To-ma-ten-sup-pe

**tomater** (Da, Nw, Sw), to-ma-ter

**tomates farcies** (Fr), to-mates far-cies

**tomatillo** (Mx), to-ma-ti-llo

**tomato** (US), to-ma-to

**tomato juice** (US), to-ma-to juice

**tomato paste** (US), to-ma-to paste

**tomato purée** (US), to-ma-to pu-rée

**tomato sauce** (US), to-ma-to sauce

**tomatsaft** (Nw), to-mat-saft

**tomatsåss** (Sw), to-mat-såss

**tomber** (Fr), tom-ber

**Tom Collins** (US), Tom Col-lins

**Tomino del Monferrato** (It), To-mi-no del Mon-fer-ra-to

**Tomme** (Fr), Tomme

**Tomme au raisin** (Fr), Tomme au rai-sin

**tomme de chèvre** (Fr), tomme de chè-vre

**tomme de Savoie** (Fr), tomme de Sa-voie

**tom sot cay** (Vt), tom sot cay

**tonfisk** (Sw), ton-fisk

**tong** (Du), tong

**tongue** (US), tongue

**tongue and blood loaf** (US), tongue and blood loaf

**tonhal** (Hu), ton-hal

**tonic water** (US), ton-ic wa-ter

**tonija** (Du), to-nija

**tonka bean** (Pg), ton-ka bean

**tonkatsu** (Jp), ton-ka-tsu

**tonnato** (It), ton-na-to

**tonnikala** (Fi), ton-ni-ka-la

**tonno** (It), ton-no

**tonno, al** (It), ton-no, al

**tonno e fagioli** (It), ton-no e fa-gio-li

**tonno sott' olio** (It), ton-no sot-t' o-lio

**toorsbee** (Tr), toor-shee

**top butt** (US), top butt

**Topf** (Gr), Topf

**Topfengolatschen** (Gr), Top-fen-go-lat-schen

**Topfenknodel** (Gr), Top-fen-kno-del

**Topfenpalatschinken** (Gr), Top-fen-pa-lat-schin-ken

**topig** (Tr), to-pig

**topinambour** (Fr), to-pi-nam-bour

**topinka** (Cz), to-pin-ka

**top loin chop** (US), top loin chop

**top of chuck** (US), top of chuck

**top round** (US), top round

**top sirloin** (US), top sir-loin

**tordi allo spiedo** (It), tor-di al-lo spie-do

**tordo** (It), tor-do

**toriganni** (Jp), to-ri-gan-ni

**tori hoban (Jp) to-ri ho-han**

**toriniku** (Jp), to-ri-ni-ku

**toriniku tatsuta age** (Jp), to-ri-ni-ku ta-tsu-ta age

**tori no nimono** (Jp), to-ri no ni-mo-no

**tori no sashimi** (Jp), to-ri no sa-shi-mi

**torkad frukt** (Sw), tor-kad frukt

**tørkage** (Da), tør-ka-ge

**tori hoban** (Jp), to-ri ho-han

**torma** (Hu), tor-ma

**tormamártás** (Hu), tor-ma-már-tás

**toronja** (Pg, Sp), to-ron-ja

**tororoimo** (Jp), to-ro-ro-i-mo

**tororo kombu** (Jp), to-ro-ro kom-bu

**tororo soba** (Jp), to-ro-ro so-ba

**torpedo** (US), tor-pe-do

**torrada** (Pg), tor-ra-da

**tørrekaker** (Nw), tør-re-kak-er

**tørret frugtsuppe** (Nw), tør-ret frugt-sup-pe

**torrfisk** (Da, Nw, Sw), torr-fisk

**torrijas** (Sp), to-rri-jas

**torrone** (It), tor-ro-ne

**torsk** (Da, Nw, Sw), torsk

**torsk med eggesaus** (Nw), torsk med eg-ge-saus

**torta** (Ph), tor-ta

**torta** (SC), tor-ta

**torta** (Pg), tor-ta

**torta** (It), tor-ta

**torta con formaggio** (It), tor-ta con for-mag-gio

**torta di macedonia di frutta** (It), tor-ta di ma-ce-do-nia di fru-tta

**torta di ricotta** (It), tor-ta di ri-cot-ta

**torta di tagliatella** (It), tor-ta di ta-glia-tel-la

**torta pasqualina** (It), tor-ta pa-squa-li-na

**Törtchen** (Gr), Tört-chen

**Torte** (Gr), Tor-te

**torte** (US), torte

**tortellata crema** (It), tor-tel-la-ta cre-ma

**tortellini** (It), tor-tel-li-ni

**tortellini alla bolognese** (It), tor-tel-li-ni al-la bo-lo-gne-se

**tortiglione** (It), tor-ti-glio-ne

**tortilha de mariscos** (Pg), tor-ti-lha de ma-ris-cos

**tortilla** (Mx), tor-ti-lla

**tortilla de huevos** (Mx), tor-ti-lla de hue-vos

**tortina** (It), tor-ti-na

**tort iz meringi** (Rs), tort iz me-rin-gi

**tortoni** (It), tor-to-ni

**tortue** (Fr), tor-tue

**toscane, à la** (Fr), tos-cane, à la

**Toscano** (It), To-sca-no

*toscatárta* (Sw), tos-ca-tår-ta

*toso* (Jp), to-so

*toss* (US), toss

*tossed salad* (US), tossed salad

*tostadas* (Mx), tos-ta-das

*tostato* (It), tos-ta-to

*tosti na jamu* (Af-Swahili), to-sti na ja-mu

*to-su* (Ch), to-su

*tōsuto* (Jp), tō-su-to

*totani* (It), to-ta-ni

*totopos* (Mx), to-to-pos

*toucinho* (Pg), tou-ci-nho

*toucinho de céu* (Pg), tou-ci-nho de céu

*toulousaine, à la* (Fr), tou-lou-saine, à la

*tourin* (Fr), tou-rin

*tournedos* (Fr), tour-ne-dos

*tournedos de veau* (Fr), tour-ne-dos de veau

*Tournedos Rossini* (Fr), Tour-ne-dos Ros-si-ni

*tourte* (Fr), tourte

*tourtelettes* (Fr), tour-te-let-tes

*toute-épice* (Fr), toute-é-pice

*tra* (Vt), tra

*tracciole d'agnello* (It), trac-cio-le d'a-gnel-lo

*trace elements* (US), trace el-e-ments

*tragacanth* (US), trag-a-canth

*tranche* (Fr), tranche

*tranebær* (Nw), tra-ne-bær

*Trappiste* (Fr), Trap-piste

*Trasch* (Gr), Trasch

*trassi* (In), tras-si

*Trauben* (Gr), Trau-ben

*travailler* (Fr), tra-vai-ller

*treacle* (GB), treac-le

*tree ear* (US), tree ear

*trefoil* (US), tre-foil

*trenette* (It), tre-net-te

*trenette con pesto* (It), tre-net-te con pes-to

*Trenton cracker* (US), Tren-ton crack-er

*trepang* (GB), trep-ang

*treska* (Cz), tres-ka

*třešně* (Cz), třeš-ně

*trešnje* (SC), tre-šnje

*trifle* (GB), tri-fle

*trifli* (Da), trif-li

*triglie* (It), tri-glie

*triglie alla livornese* (It), tri-glie al-la li-vor-ne-se

*trigo* (Sp), tri-go

*trigo negro* (Sp), tri-go ne-gro

*tripa* (Pg), tri-pa

*tripe* (US), tripe

*tripes à la mode de Caen* (Fr), tripes à la mode de Caen

*triple-crème* (Fr), tri-ple-crème

*triple sec* (US), tri-ple sec

*trippa alla bolognese* (It), trip-pa al-la bo-lo-gne-se

*trippa alla fiorentina* (It), trip-pa al-la fio-ren-ti-na

*trippa al sugo* (It), trip-pa al su-go

*triticale flour* (US), trit-i-ca-le flour

*Trockenbeerenauslese* (Gr),

Tro-cken-bee-re-naus-le-
se

*tronçons de homard* (Fr),
tron-çons de ho-mard

*trota* (It), tro-ta

*trota salmonata* (It), tro-ta
sal-mo-na-ta

*trotter* (GB), trot-ter

*trout* (US), trout

*trouvillaise* (Fr), trou-vi-
llaise

*trucha* (Sp), tru-cha

*trufa* (Pg), tru-fa

*truffe* (Fr), truffe

*Trüffel* (Gr), Trüf-fel

*truffle* (US), truf-fle

*truite* (Fr), truite

*truite au bleu de meunière*
(Fr), truite au bleu de
meu-nière

*truite saumonée* (Fr), truite
sau-mo-née

*trunzo* (It), trun-zo

*truskawki* (Po), trus-kaw-ki

*truss* (US), truss

*truta* (Pg), tru-ta

*Truthahn* (Gr), Trut-hahn

*Truthahn mit Reis gra-
tiniert* (Gr), Trut-hahn mit
Reis gra-ti-niert

*tryptophan* (US), tryp-to-
phan

*tsai* (Gk), tsa-i

*tsài tāng* (Ch), tsài tāng

*tsán-dòu* (Ch), tsán-dòu

*tsău-méi* (Ch), tsău-méi

*tsetov* (Tr), tset-ov

*tsetov gangar* (Tr), tset-ov
gan-gar

*tsetov leetsk* (Tr), tset-ov
leetsk

*tsetov tzavari yeghintz* (Tr),
tset-ov tza-var-i ye-ghintz

*tsgnaganch* (Tr), tsg-na-
ganch

*tsgnaganchi leetsk* (Tr), tsg-
na-ganch-i leetsk

*tsgnaganchi meechoog* (Tr),
tsg-na-ganch-i mee-choog

*tsoog* (Tr), tsoog

*tsoogi aghtsan* (Tr), tsoog-i
agh-tsan

*tsù* (Ch), tsù

*tsukemono* (Jp), tsu-ke-mo-
no

*tsukimi udon* (Jp), tsu-ki-mi
u-don

*tsukudani* (Jp), tsu-ku-da-ni

*tsumamimono* (Jp), tsu-ma-
mi-mo-no

*tswèi-pí jī* (Ch), tswèi-pí jī

*tsyplyata tabaka* (Rs), tsy-
plya-ta ta-ba-ka

*tubettini* (It), tu-bet-ti-ni

*tufaa* (Af-Swahili), tu-fa-a

*tuffaaha* (Ar), tuf-faa-ha

*tuile* (Fr), tuile

*tükör tojás* (Hu), tü-kör to-jás

*tulband* (Du), tul-band

*tulipe* (Fr), tu-lipe

*tulsee* (Ia), tul-see

*tum* (Ar), tum

*tuna* (US), tu-na

*tuna* (Mx), tu-na

*tuňák* (Cz), tu-ňák

*tunge* (Da), tung-e

*tunge* (Nw), tung-e

*tungeflyndre* (Nw), tung-e-
flyn-dre

*tungule* (Af-Swahili), tu-ngu-le

*tunjevina* (SC), tu-nje-vi-na

*Tunke* (Gr), Tun-ke

*tunnied veal* (GB), tun-nied veal

*tunny* (GB), tun-ny

*tuoremehu* (Fi), tu-o-re-me-hu

*tuorlo d'uova* (It), tuor-lo d'uo-va

*tur* (Ia), tur

*turbinado sugar* (US), tur-bin-a-do sug-ar

*turbot* (US), tur-bot

*turbot poché hollandaise* (Fr), tur-bot po-ché hol-lan-daise

*tur dal* (Ia), tur dal

*turkey* (US), tur-key

*Turkish Delight* (US), Turk-ish De-light

*türlü* (Tr), tür-lü

*turmeric* (US), tur-mer-ic

*turn* (US), turn

*turnedo* (It), tur-ne-do

*turnip* (US), tur-nip

*turnip greens* (US), tur-nip greens

*turnip-root celery* (US), tur-nip-root cel-er-y

*turnip-rooted cabbage* (US), tur-nip-root-ed cab-bage

*turnover* (US), turn-o-ver

*túrós* (Hu), tú-rós

*túrós gombóc* (Hu), tú-rós gom-bóc

*túróscsusza* (Hu), tú-rósc-su-sza

*turska* (Fi), turs-ka

*turşu* (Tr), tur-şu

*turtle* (US), tur-tle

*tutmaj* (Tr), tut-maj

*tutoo* (Tr), tu-too

*tutoo aboor* (Tr), tu-too a-boor

*tutoom* (Tr), tu-toom

*tutoomi bahadzo* (Tr), tu-toomi ba-ha-dzo

*tutoomi leetsk* (Tr), tu-toomi leetsk

*tutti-frutti* (It), tut-ti-frut-ti

*tutvash* (Tr), tut-vash

*tuz* (Tr), tuz

*tvaroh* (Cz), tva-roh

*tvorog* (Rs), tvo-rog

*tvrda jaja* (SC), tvr-da ja-ja

*twaalfuurtje* (Du), twaal-fuur-tje

*twelfth-night cake* (GB), twelfth-night cake

*tyán-gwā* (Ch), tyán-gwā

*tyán chéng chá* (Ch), tyán chéng chá

*tyán swān gú lău ròu* (Ch), tyán swān gú lău ròu

*tyán swān yú* (Ch), tyán swān yú

*Tybo* (Da), Ty-bo

*tykmælk* (Da), tyk-mælk

*tyrolienne* (Fr), ty-ro-lienne

*tyrosine* (US), ty-ro-sine

*tyttebær* (Da), tytt-e-bær

*tyttebær* (Nw), tytt-e-bær

*tzavar* (Tr), tza-var

*tzavari aghtsan* (Tr), tza-vari agh-tsan

*tzavari yeghintz* (Tr), tza-vari ye-ghintz

*tzimmes* (Jw), tzim-mes

*tzoren* (Tr), tzo-ren

# U

*überbacken* (Gr), ü-ber-bac-ken

*uborka* (Hu), u-bor-ka

*uborkamártás* (Hu), u-bor-ka-már-tás

*uborkasaláta* (Hu), u-bor-ka-sa-lá-ta

*uccèlletti* (It), uc-cèl-let-ti

*uccèlletti scappati* (It), uc-cèl-let-ti scap-pa-ti

*uccèlli* (It), uc-cèl-li

*udang* (In), u-dang

*udang asam manis* (In), u-dang a-sam ma-nis

*udang goreng* (In), u-dang go-reng

*udang karang* (In), u-dang ka-rang

*udang kerie* (In), u-dang ke-rie

*udang pindang ketjap* (In), u-dang pin-dang ket-jap

*udo* (Jp), u-do

*udon* (Jp), u-don

*udonsuki* (Jp), u-don-su-ki

*udruk* (Ia), ud-ruk

*ugli fruit* (Cb), ug-li fruit

*ugnsbakat* (Sw), ugns-ba-kat

*ugnspannkaka* (Sw), ugns-pann-ka-ka

*ugnsstekt* (Sw), ugns-ste-kt

*úboř* (Cz), ú-hoř

*uien* (Du), ui-en

*uitsmijter* (Du), uits-mij-ter

*uji* (Af-Swahili), u-ji

*ukad* (Ia), u-kad

*ukba* (Rs), u-kha

*uku* (Pl), u-ku

*ukwaju* (Af-Swahili), u-kwa-ju

*ulje* (SC), u-lje

*ulu-ulu* (Pl), u-lu-u-lu

*umani* (Jp), u-ma-ni

*um ar-rubiyam* (Ar), um ar-ru-bi-yam

*umchur* (Ia), um-chur

*umeboshi* (Jp), u-me-bo-shi

*ume maki* (Jp), u-me ma-ki

*umeshu* (Jp), u-me-shu

*umewan* (Jp), u-me-wan

*umido* (It), um-i-do

*umido di coniglio* (It), u-mi-do di co-ni-glio

*umintas* (Sp), u-min-tas

*umm ali* (Ar), umm a-li

*unagi* (Jp), u-na-gi

*unagi domburi* (Jp), u-na-gi dom-bu-ri

*unagi maki* (Jp), u-na-gi ma-ki

**unbleached flour** (US), un-bleach-ed flour

**unday** (Ia), un-day

**unday brinjal** (Ia), un-day brin-jal

**unelmapannukakku** (Fi), u-nel-ma-pan-nu-kak-ku

**uni** (Jp), u-ni

**univalve** (US), un-i-valve

**unleavened** (US), un-leav-ened

**unpolished rice** (US), un-pol-ished rice

**unsaturated fat** (US), un-sat-u-rat-ed fat

**unsaturated fatty acid** (US), un-sat-u-rat-ed fat-ty ac-id

**uova** (It), uo-va

**uova affogate** (It), uo-va af-fo-ga-te

**uova affogate con acciughe** (It), uo-va af-fo-ga-te con ac-ciu-ghe

**uova al prosciutto** (It), uo-va al pro-sciut-to

**uova bollite** (It), uo-va bol-li-te

**uova fritte** (It), uo-va frit-te

**uova morene** (It), uo-va mo-re-ne

**uova piccante** (It), uo-va pic-can-te

**uova sode con tonno** (It), uo-va so-de con ton-no

**uova strapazzate** (It), uova stra-paz-za-te

**upside-down cake** (US), up-side-down cake

**urad** (Ia), u-rad

**urap** (In), u-rap

**urme** (SC), ur-me

**ürübús** (Hu), ü-rü-hús

**ushio wan** (Jp), u-shi-o wan

**ushki** (Rs), ush-ki

**ústřice** (Cz), úst-ři-ce

**ustritsy s ikroi** (Rs), us-tri-tsy s i-kroi

**usukuchi shoyu** (Jp), u-su-ku-chi sho-yu

**uszka** (Po), usz-ka

**uthappam** (In), u-thap-pam

**utrunj** (Ar), u-trunj

**uuden vuoden malja** (Fi), uu-den vu-o-den mal-ja

**uunijuustoa** (Fi), uu-ni-juus-to-a

**uuni riisipuuro** (Fi), uu-ni rii-si-puu-ro

**uunissa paistettu** (Fi), uu-nis-sa pais-tet-tu

**uunissa paistettu perunakakkuja** (Fi), uu-nis-sa pais-tet-tu pe-ru-na-kak-ku-ja

**uva** (Pg, Sp), u-va

**uva** (It), u-va

**uva passa** (It), u-va pas-sa

**uva spina** (It), u-va spi-na

**uyoga** (Af-Swahili), u-yo-ga

**uzená šunka** (Cz), u-ze-ná šun-ka

**uzený jazyk** (Cz), u-ze-ný ja-zyk

**üzüm** (Tr), ü-züm

**uzura** (Jp), u-zu-ra

**uzura no tamago** (Jp), u-zu-ra no ta-ma-go

# 𝒱

*vaca* (Pg), va-ca
*vaca cozida* (Pg), va-ca co-zi-da
*vaca estufada* (Pg), va-ca es-tu-fa-da
*vaca guisada* (Pg), va-ca gui-sa-da
*vacherin* (Fr), va-che-rin
*Vacherin Mont d'Or* (Fr-Swiss), Va-che-rin Mont d'Or
*vad* (Hu), vad
*vadas* (Ia), va-das
*vadelmia* (Fi), va-del-mi-a
*våfflor* (Sw), våff-lor
*vafler* (Nw), vaf-ler
*vaflya* (Rs), vaf-lya
*vahva juusto* (Fi), vah-va juus-to
*vainilla* (Sp), va-i-ni-lla
*vaj* (Hu), vaj
*vajas pogácsa* (Hu), va-jas po-gá-csa
*vajbab* (Hu), vaj-bab
*vajgaluska* (Hu), vaj-ga-lus-ka
*vaktel* (Sw), vak-tel
*val* (Ia), val
*Valencay* (Fr), Va-len-cay

*valenciano* (Mx), va-len-ci-a-no
*Valencia orange* (US), Va-len-ci-a or-ange
*Valencienne, à la* (Fr), Val-en-ci-enne, à la
*valine* (US), val-ine
*valkosipuli* (Fi), val-ko-si-pu-li
*valkosipulisilakka* (Fi), val-ko-si-pu-li-si-lak-ka
*valkoviiniä* (Fi), val-ko-vii-ni-ä
*valnødkage* (Da), val-nød-ka-ge
*valnødromkager* (Da), val-nød-rom-ka-ger
*valnötter* (Sw), val-nött-er
*valnøtter* (Nw), val-nøtt-er
*Valois, sauce* (Fr), Val-ois, sauce
*valpolicella* (It), val-po-li-cel-la
*vanaspati* (Ia), va-nas-pa-ti
*vand* (Da), vand
*vaniglia* (It), va-ni-glia
*vanilia fagylalt* (Hu), va-ni-li-a fagy-lalt
*vaniljajäätelö* (Fi), va-nil-ja-jää-te-lö

*vanilje* (Nw), va-nil-je

*vaniljepudding* (Nw), va-nil-je-pud-ding

*vaniljglass* (Sw), va-nilj-glass

*vaniljsås* (Sw), va-nilj-sås

*vanilla* (US), va-nil-la

*vanilla extract* (US), va-nil-la ex-tract

*vanilla sugar* (US), va-nil-la sug-ar

*vanille* (Fr), va-nille

*Vanilleeis* (Gr), Va-nil-le-eis

*vanilleijs* (Du), va-nil-le-ijs

*vanillin* (US), va-nil-lin

*vann* (Nw), vann

*vannbakkelse* (Nw), vann-bak-kel-se

*vannmelon* (Nw), vann-me-lon

*vanocka* (Cz), va-noc-ka

*vapeur, à la* (Fr), va-peur, à la

*varak* (Ia), va-rak

*vařené* (Cz), va-ře-né

*vařené brambory* (Cz), va-ře-né bram-bo-ry

*vařené hovřzi maso* (Cz), va-ře-né ho-vě-zi ma-so

*vařené vejce* (Cz), va-ře-né vej-ce

*vareniki* (Rs), va-re-ni-ki

*varerdrikke* (Nw), var-er-drikk-e

*varié* (Fr), va-ri-é

*variety meats* (US), va-ri-e-ty meats

*varkenskarbonaden* (Du), var-kens-kar-bo-na-den

*varkenskotelet* (Du), var-kens-ko-te-let

*varkensvlees* (Du), var-kens-vlees

*varm choklad* (Sw), varm chok-lad

*varme pølser* (Nw), var-me pøl-ser

*varm krabbsmörgåss* (Sw), varm krabb-smör-gåss

*vartaassa* (Fi), var-taas-sa

*vasikanleike* (Fi), va-si-kan-le-i-ke

*vasikanliha* (Fi), va-si-kan-li-ha

*vasilopeta* (Gk), va-si-lo-pe-ta

*västkustsallad* (Sw), väst-kust-sal-lad

*vatapá* (Pg), va-ta-pá

*vath ka salun* (Ia), vath ka sa-lun

*vatkattu marjapuuro* (Fi), vat-kat-tu mar-ja-puu-ro

*vatrushki* (Rs), va-trush-ki

*vatten* (Sw), vat-ten

*vattenglass* (Sw), vat-ten-glass

*vaxbönor* (Sw), vax-bö-nor

*veado* (Pg), ve-a-do

*veal* (US), veal

*Veal Cordon Bleu* (US), Veal Cor-don Bleu

*veal flory* (Sc), veal flo-ry

*veal Orloff* (US), veal Or-loff

*veal parmigiana* (US), veal par-mi-gia-na

*veau* (Fr), veau

*veau thonné* (Fr), veau thon-né

*vegetable extract* (US), veg-e-ta-ble ex-tract

*vegetable marrow* (GB), veg-e-ta-ble mar-row

*vegetable oyster* (US), veg-e-ta-ble oys-ter

*vegetable pear* (US), veg-e-ta-ble pear

*vegetarian* (US), veg-e-tar-i-an

*vegetarian duck* (US), veg-e-tar-i-an duck

*vegyes saláta* (Hu), ve-gyes sa-lá-ta

*vejce* (Cz), vej-ce

*vejce na měkko* (Cz), vej-ce na měk-ko

*vejce na tvrdo* (Cz), vej-ce na tvr-do

*vellutata* (It), vel-lu-ta-ta

*velouté, sauce* (Fr), ve-lou-té, sauce

*velouté de légumes* (Fr), ve-lou-té de lé-gumes

*velouté de volaille* (Fr), ve-lou-té de vo-laille

*velstekt* (Nw), vel-stekt

*velveting* (US), vel-vet-ing

*venado* (Sp), ve-na-do

*venaison* (Fr), ve-nai-son

*vengi bath* (Ia), ven-gi bath

*venison* (US), ven-i-son

*vénitienne, sauce* (Fr), vé-ni-tienne, sauce

*venkel* (Du), ven-kel

*vepřová pečeně* (Cz), vep-řo-vá pe-če-ně

*vepřové maso* (Cz), vep-řo-vé ma-so

*verbena* (US), ver-be-na

*verde* (It, Mx), ver-de

*verdura* (It), ver-du-ra

*verduras* (Pg), ver-du-ras

*verdure cotte* (It), ver-du-re cot-te

*verdurette, sauce* (Fr), ver-dur-ette, sauce

*véres hurka* (Hu), vé-res hur-ka

*verikoka* (Gk), ve-ri-ko-ka

*veriohukaiset* (Fi), ve-ri-o-hu-ka-i-set

*veripalttu* (Fi), ve-ri-palt-tu

*verjuice* (US), ver-juice

*verlorene Eier* (Gr), ver-lo-re-ne Ei-er

*vermicelli* (It), ver-mi-cel-li

*Véronique* (Fr), Vé-ro-nique

*verschieden* (Gr), ver-schie-den

*verte, sauce* (Fr), verte, sauce

*vert-pré, au* (Fr), vert-pré, au

*very-low-sodium* (US), ver-y-low-so-di-um

*verza* (It), ver-za

*verzata di riso* (It), ver-za-ta di ri-so

*vešalica* (SC), ve-ša-li-ca

*vese gombával* (Hu), ve-se gom-bá-val

*vesék* (Hu), ve-sék

*vese velö tojással* (Hu), ve-se ve-lö to-jás-sal

*vetchina* (Rs), vet-chi-na

*vettä* (Fi), vet-tä

*viande* (Fr), vi-ande

*viande froides* (Fr), vi-ande froides

*viazi* (Af-Swahili), vi-a-zi

*viazi vya kuchemsha* (Af-Swahili), vi-a-zi vya ku-chem-sha

*viazi vya kukaanga* (Af-Swahili), vi-a-zi vya ku-kaa-nga

*viazi vya kuvuruga* (Af-Swahili), vi-a-zi vya ku-vu-ru-ga

*viazi vitamu* (Af-Swahili), vi-a-zi vi-ta-mu

*viceroy's dessert* (Mx), vi-ce-roy's des-sert

*vichy* (Fr), vi-chy

*vichyssoise* (US), vi-chy-ssoise

*vichyssoise à la Russe* (Fr), vi-chy-ssoise à la Russe

*Vichy water* (US), Vi-chy wa-ter

*Victoria, à la* (Fr), Vic-to-ria, à la

*Victoria sandwich* (GB), Vic-to-ria sand-wich

*Vidalia onions* (US), Vi-dal-ia on-ions

*vídeňský řízek* (Cz), ví-deň-ský ři-zek

*Viennoise, à la* (Fr), Vi-en-noise, à la

*Vierfrüchtkuchen* (Gr), Vier-frücht-ku-chen

*viikunoita* (Fi), vii-ku-no-i-ta

*viili* (Fi), vii-li

*viinimarjoja* (Fi), vii-ni-mar-jo-ja

*viinirypäleitä* (Fi), vii-ni-ry-pä-lei-tä

*vijgen* (Du), vij-gen

*vild bönsfåsgel* (Du), vild höns-fås-gel

*Villalón* (Sp), Vi-lla-lón

*villeroi, sauce* (Fr), ville-roi, sauce

*vin* (Sw), vin

*vin, au* (Fr), vin, au

*vinäger* (Sw), vi-nä-ger

*vinagre* (Sp), vi-na-gre

*vinagre* (Pg), vi-na-gre

*vinaigre* (Fr), vi-nai-gre

*vinaigré* (Fr), vi-nai-gré

*vinaigrette* (Fr), vi-nai-grette

*vinaigrette huile de noix* (Fr), vi-nai-grette huile de noix

*vinbär* (Sw), vin-bär

*vin blanc* (Fr), vin blanc

*Vincent, sauce* (Fr), Vin-cent, sauce

*vindaloo* (Ia), vin-da-loo

*vindruer* (Da), vin-druer

*vindruva* (Du), vin-dru-va

*vindruvor* (Sw), vin-dru-vor

*vinegar* (US), vin-e-gar

*vine leaves* (US), vine leaves

*vinete cu carne* (SC), vi-ne-te cu car-ne

*vinho branco* (Pg), vin-ho bran-co

*vinho do Porto* (Pg), vin-ho do Por-to

*vinho tinto* (Pg), vin-ho tin-to

*vin hvit* (Nw), vin hvit

*vino* (Cz), vi-no

*vino bianco* (It), vi-no bian-co

*vino da tavola* (It), vi-no da ta-vo-la

*vino rosso* (It), vi-no ros-so

*vino secco* (It), vi-no sec-co

*vin rød* (Nw), vin rød

*vin rouge* (Fr), vin rouge

*vinsuppe* (Nw), vin-sup-pe

*vintage* (US), vin-tage

*violet* (US), vi-o-let

*Virginia ham* (US), Vir-gin-ia ham

*virgin olive oil* (US), vir-gin ol-ive oil

*Viroflay, à la* (Fr), Vi-ro-flay, à la

*vis* (Du), vis

*visciola* (It), vis-ci-o-la

*višisoaz* (SC), vi-ši-so-az

*viskoekjes* (Du), vis-koek-jes

*višně* (Cz, Tr), viš-ně

*vişne suyu* (Tr), viş-ne su-yu

*vispgrädde* (Sw), visp-gräd-de

*vitamin* (US), vit-a-min

*vitamin A* (US), vit-a-min A

*vitamin B complex* (US), vit-a-min B com-plex

*vitamin B1* (US), vit-a-min B1

*vitamin B2* (US), vit-a-min B2

*vitamin B3* (US), vit-a-min B3

*vitamin B5* (US), vit-a-min B5

*vitamin B6* (US), vit-a-min B6

*vitamin B12* (US), vit-a-min B12

*vitamin C* (US), vit-a-min C

*vitamin D* (US), vit-a-min D

*vitamin E* (US), vit-a-min E

*vitamini* (Rs), vi-ta-mi-ni

*vitamin K* (US), vit-a-min K

*Vitaminmangel* (Gr), Vi-ta-min-man-gel

*vitaminy* (Cz), vi-ta-mi-ny

*vitela* (Pg), vi-te-la

*vitello* (It), vi-tel-lo

*vitello alla Milanese* (It), vi-tel-lo al-la Mi-la-nese

*vitello tonnato* (It), vi-tel-lo ton-na-to

*vitkåslsoppa med kroppkakor* (Sw), vit-kåsl-sop-pa med kropp-ka-kor

*vitling* (Sw), vit-ling

*vitlök* (Sw), vit-lök

*vitsåss* (Sw), vit-såss

*vitt bröd* (Sw), vitt bröd

*vitt vin* (Sw), vitt vin

*vitunguu* (Af-Swahili), vi-tu-ngu-u

*viz* (Hu), viz

*vla* (Du), vla

*vlašské ořechy* (Cz), vlaš-ské o-ře-chy

*vlees* (Du), vlees

*vleespannekoekjes* (Du), vlees-pan-ne-koek-jes

*vlees voor de boterham* (Du), vlees voor de bo-ter-ham

*vleet* (Du), vleet

*voće* (SC), vo-će

*voćni sok* (SC), vo-ćni sok

*voda* (Cz), vo-da

*vodka* (Rs), vod-ka

*vodu sa ledom* (SC), vo-du sa le-dom

*vohveli* (Fi), voh-ve-li

*voi* (Fi), voi

*voileipä* (Fi), voi-le-i-pä

*voileipäpöytä* (Fi), voi-le-i-pä-pö-y-tä

*voimakkaasti maustettua* (Fi), voi-mak-kaas-ti ma-us-tet-tu-a

*volaille* (Fr), vo-laille

*volaille au vinaigre* (Fr), vo-laille au vin-ai-gre

*vol-au-vent* (Fr), vol-au-vent

*vol-au-vent de ris de veau* (Fr), vol-au-vent de ris de veau

*volos* (Gr), vo-los

*vongole* (It), von-go-le

*vongole alla marinara* (It), von-go-le al-la ma-ri-na-ra

*vongole ripiene al forno* (It), von-go-le ri-pie-ne al for-no

*voorgerechten* (Du), voor-ge-rech-ten

*vörösbort* (Hu), vö-rös-bort

*Vorspeisen* (Gr), Vor-spei-sen

*vörtbröd* (Sw), vört-bröd

*vørterkake* (Nw), vør-ter-ka-ke

*vruća čokolada* (SC), vru-ća čo-ko-la-da

*vrucht* (Du), vrucht

*vruchtensap* (Du), vruch-ten-sap

*vruchtentaart* (Du), vruch-ten-taart

*vutiro* (Gk), vu-ti-ro

# W

*Wachenheimer* (Gr), Wa-chen-hei-mer

*Wachsbohnen* (Gr), Wachs-boh-nen

*Wachteln* (Gr), Wach-teln

*wadas* (Ia), wa-das

*wadjid* (In), wad-jid

*wafels* (Du), wa-fels

*wafer* (US), wa-fer

*Waffeln* (Gr), Waf-feln

*waffle* (US), waf-fle

*wagashi* (Jp), wa-ga-shi

*waha dori teriyaki* (Jp), wa-ha do-ri te-ri-ya-ki

*Wähen* (Gr-Swiss), Wäh-en

*wahoo* (US), wa-hoo

*wain* (Ml), wain

*wajik* (In), wa-jik

*wakame* (Jp), wa-ka-me

*wakasagi* (Jp), wa-ka-sa-gi

*wakasagi no furai* (Jp), wa-ka-sa-gi no fu-ra-i

*wakegi* (Jp), wa-ke-gi

*Waldegerling* (Gr), Wal-de-ger-ling

*Waldmeister* (Gr), Wald-mei-ster

*Waldmeisterbraten* (Gr), Wald-mei-ster-bra-ten

*Waldorfsalad* (US), Wal-dorf sal-ad

*Waldschnepfe* (Gr), Wald-schnep-fe

*Walewska, à la* (Fr), Wa-lew-ska, à la

*wali* (Af-Swahili), wa-li

*Waller gebacken* (Gr), Wal-ler ge-bac-ken

*walleye* (US), wall-eye

*walnoot* (Du), wal-noot

*Walnuss* (Gr), wal-nuss

*walnut* (US), wal-nut

*walnut oil* (US), wal-nut oil

*wampi* (Th), wam-pi

*wān dòu* (Ch), wān dòu

*wanilia* (Po), wa-nil-ia

*warabi* (Jp), wa-ra-bi

*warak al gar* (Ar), wa-rak al gar

*warak inib* (Ar), wa-rak i-nib

*warak inib mihshee* (Ar), wa-rak i-nib mih-shee

*warm, to* (US), warm, to

*Warmbier* (Gr), Warm-bier

*Warme Wurstspeisen* (Gr), Warme Wurst-spei-sen

*wasabi* (Jp), wa-sa-bi

*wash* (US), wash

*wassail* (GB), was-sail

*Wasser* (Gr), Wa-sser

*Wassermelone* (Gr), Wa-sser-mel-o-ne

*Wasserteig* (Gr), Wa-sser-teig

*water* (US), wa-ter

*water biscuit* (US), wa-ter bis-cuit

*water chestnut* (US), wa-ter chest-nut

*waterchocolade* (Du), wa-ter-cho-co-la-de

*watercress* (US), wa-ter-cress

*watercress sauce* (US), wa-ter-cress sauce

*water ice* (US), wa-ter ice

*watermelon* (US), wa-ter-mel-on

*waterzootje* (Bl), wa-ter-zoo-tje

*waterzootje de poulet* (Bl), wa-ter-zoo-tje de pou-let

*watróbka* (Po), wa-trób-ka

*wau* (Ch), wau

*wax bean* (US), wax bean

*weakfish* (US), weak-fish

*Weckklösse* (Gr), Weck-klös-se

*wedding cake* (US), wed-ding cake

*wěi* (Ch), wěi

*weichgekochte Eier* (Gr), weich-ge-koch-te Ei-er

*wèi-jīng* (Ch), wèi-jīng

*Wein* (Gr), Wein

*Weinbergschnecken* (Gr), Wein-berg-schne-cken

*Weinbrand* (Gr), Wein-brand

*Weinkaltschale* (Gr), Wein-kalt-scha-le

*Weinkäse* (Gr), Wein-kä-se

*Weinkraut* (Gr), Wein-kraut

*Weinsuppe* (Gr), Wein-sup-pe

*Weintraube* (Gr), Wein-trau-be

*Weissbrot* (Gr), Weiss-brot

**Weisse Bohnen** (Gr), Weisse Boh-nen

**Weisse Bratwürste** (Gr), Weisse Brat-würste

**Weisse Rüben** (Gr), Weisse Rü-ben

**Weisse Sosse** (Gr), Weisse Sos-se

**Weissfisch** (Gr), Weiss-fisch

**Weisskäse** (Gr), Weiss-kä-se

**Weisskohl** (Gr), Weiss-kohl

**Weisslacker** (Gr), Weiss-lacker

**Weissrüben** (Gr), Weiss-rüben

**Weisswein** (Gr), Weiss-wein

**Weisswürste** (Gr), Weiss-würste

**Weizen** (Gr), Wei-zen

**Welsh border tart** (GB), Welsh bor-der tart

**Welsh rabbit** (GB), Welsh rab-bit

**Wensleydale cheese** (GB), Wens-ley-dale cheese

**Werderkäse** (Gr), Wer-der-kä-se

**west coast halibut royal** (Ca), west coast hal-i-but roy-al

**western** (US), west-ern

**Westfälischer frischer Obstkuchen** (Gr), West-fä-li-scher fri-scher Obst-ku-chen

**Westfälischer Schinken** (Gr), West-fä-li-scher Schin-ken

**whale** (US), whale

**wheat** (US), wheat

**wheat flour** (US), wheat flour

**wheat germ** (US), wheat germ

**whelk** (GB), whelk

**whey** (US), whey

**whip** (US), whip

**whip, to** (US), whip, to

**whipped cream** (US), whipped cream

**whipped topping** (US), whipped top-ping

**whipping cream** (US), whipping cream

**whisk, to** (US), whisk, to

**whitebait** (US), white-bait

**white basil** (Ia), white ba-sil

**white bean** (US), white bean

**white butter sauce** (US), white but-ter sauce

**white chocolate** (US), white choc-o-late

**whitefish** (US), white-fish

**white flour** (US), white flour

**white milk sauce** (US), white milk sauce

**white pepper** (US), white pepper

**white pudding** (GB), white pud-ding

**white sauce** (US), white sauce

**white vinegar** (US), white vin-e-gar

**white walnut** (US), white wal-nut

**white wine sauce** (US), white wine sauce

**whiting** (US), whit-ing

**whole grain** (US), whole grain

**wholemeal flour** (GB), whole-meal flour

*whole wheat flour* (US), whole wheat flour

*whortleberry* (GB), whor-tle-ber-ry

*widjen* (In), wid-jen

*wiener* (US), wie-ner

*Wiener Backhendl* (Gr), Wie-ner Back-hen-dl

*wienerbröd* (Sw), wie-ner-bröd

*wienerbrød* (Da, Nw), wie-ner-brød

*Wiener Krapfen* (Gr), Wie-ner Kra-pfen

*wienerleipä* (Fi), wie-ner-le-i-pä

*Wiener Rostbraten* (Gr), Wie-ner Rost-bra-ten

*Wiener Schlagobers* (Gr-Austria), Wie-ner Schla-go-bers

*Wiener Schnitzel* (Gr), Wie-ner Schnit-zel

*Wienerwurst* (Gr), Wie-ner-wurst

*wieprzowina* (Po), wiepr-zo-wi-na

*wijn* (Du), wijn

*wild* (Du), wild

*Wild* (Gr), Wild

*Wildbretpastete* (Gr), Wild-bret-pas-te-te

*wilde appel* (Du), wil-de ap-pel

*wilde eend met sinaasappel* (Du), wil-de eend met si-naas-ap-pel

*Wildente* (Gr), Wild-en-te

*wildfowl* (US), wild-fowl

*Wildgeflügel* (Gr), Wild-ge-flü-gel

*wild rice* (US), wild rice

*Wildschweinbraten* (Gr), Wild-schwein-bra-ten

*wild strawberry* (GB), wild straw-ber-ry

*Windbeutel* (Gr), Wind-beu-tel

*wine* (US), wine

*wineberry* (US), wine-ber-ry

*wine vinegar* (US), wine vin-e-gar

*winkle* (GB), win-kle

*wino biale* (Po), wi-no bia-le

*wino czerwone* (Po), wi-no czer-wo-ne

*winogrona* (Po), wi-no-gro-na

*wintergreen* (US), win-ter-green

*Winterkohl* (Gr), Win-ter-kohl

*winter melon* (US), win-ter mel-on

*winter pear* (US), win-ter pear

*winter savory* (US), win-ter sa-vor-y

*Wirsingkohl* (Gr), Wir-sing-kohl

*wiśnie* (Po), wiś-nie

*witlof* (Du), wit-lof

*witte bonen* (Du), wit-te bo-nen

*wittebrood* (Du), wit-te-brood

*witte kool* (Du), wit-te kool

*witte wijn* (Du), wit-te wijn

*wok* (US), wok

**Wolfsbarsch** (Gr), Wolfs-barsch
**wonton** (US), won-ton
**wonton wrappers** (US), won-ton wrap-pers
**wood ear** (US), wood ear
**woodcock** (GB), wood-cock
**woodruff** (US), wood-ruff
**Worcester sauce** (GB), Worces-ter sauce
**Worcestershire sauce** (GB), Worces-ter-shire sauce
**wormseed** (US), worm-seed
**worst** (Du), worst
**wortel** (In), wor-tel
**worteltjes** (Du), wor-tel-tjes

**wrasse** (US), wrasse
**wú-hwā-gwo** (Ch), wú-hwā-gwo
**wū lúng chà** (Ch), wū lúng chà
**Wurst** (Gr), Wurst
**Wurstbrot** (Gr), Wurst-brot
**Würstchen** (Gr), Würst-chen
**Würze** (Gr), Würze
**Wurzeln** (Gr), Wur-zeln
**wǔ-syāng jī** (Ch), wǔ-syāng jī
**wǔ-syāng-lyáu** (Ch), wǔ-syāng-lyàu
**wǔ-syāng pái-gǔ** (Ch), wǔ-syāng pái-gǔ
**wuz** (Ar), wuz

# X

**xacutti** (Pg), xa-cut-ti
**xalota** (Pg), xa-lo-ta
**xamfina** (Sp), xam-fi-na
**xanthide** (US), xan-thide
**xapoipa** (Sp), xa-po-i-pa
**xarel-lo** (Sp), xa-rel-lo
**xarope** (Pg), xa-ro-pe
**Xavier** (Fr), Xa-vi-er
**xérès, au** (Fr), xé-rès, au
**xia** (Ch), xia
**xiangjiao** (Ch), xiang-jiao

**xiangyóu** (Ch), xiang-yóu
**ximénia** (Fr), xi-mé-ni-a
**xìngrén dòufù** (Ch), xìng-rén dòu-fù
**xingzi** (Ch), xing-zi
**xinxin de galinha** (Pg), xin-xin de ga-li-nha
**xoconoxtles** (Mx), xo-co-nox-tles
**xylitol** (US), xy-li-tol

# Y

**yabloki** (Rs), ya-blo-ki
**yabloki sup** (Rs), ya-blo-ki
  sup
**yā-dz** (Ch), yā-dz
**yā-dz dàn-jywan** (Ch), yā-dz
  dàn-jywan
**yagodni sup** (Rs), ya-god-ni
  sup
**yabni** (Tr), yah-ni
**yai la kukaanga** (Af-Swahili), ya-i la ku-ka-a-nga
**yaita** (Jp), ya-i-ta
**yaitsa** (Rs), ya-i-tsa
**yā jyàng** (Ch), yā jyàng
**yakbni** (Ia), yakh-ni
**yakidofu** (Jp), ya-ki-do-fu
**yaki hamaguri** (Jp), ya-ki ha-
  ma-gu-ri
**yaki ika** (Jp), ya-ki i-ka
**yakimono** (Jp), ya-ki-mo-no
**yakisoba** (Jp), ya-ki-so-ba
**yakitori** (Jp), ya-ki-to-ri
**yakizakana** (Jp), ya-ki-za-ka-
  na
**yakumi** (Jp), ya-ku-mi
**yalanci** (Tr), ya-lan-ci
**yalas çorbası** (Tr), ya-las çor-
  ba-sı

**yam** (Th), yam
**yams** (US), yams
**yán** (Ch), yán
**yáng-bái-tsài** (Ch), yáng-bái-
  tsài
**yáng-ròu** (Ch), yáng-ròu
**yáng-tsài-bwār** (Ch), yáng-
  tsài-hwār
**yáng-tsūng** (Ch), yáng-tsūng
**yān-mài** (Ch), yān-mài
**yansoon** (Ar), yan-soon
**yān wō** (Ch), yān wō
**yaourt** (Fr), ya-ourt
**yaout** (Rs), ya-out
**yaprek dolmasi** (Tr), ya-prek
  dol-ma-si
**yard-long bean** (US), yard-
  long bean
**yarrow** (US), yar-row
**yasai** (Jp), ya-sa-i
**yasai sūpu** (Jp), ya-sa-i sū-pu
**yassa** (Af), yas-sa
**yassa au poulet** (Af), yas-sa
  au pou-let
**yataklete kilkil** (Af), ya-tak-
  le-te kil-kil
**yāu-dòu** (Ch), yāu-dòu
**yāu-dz** (Ch), yāu-dz
**yāu-gwo** (Ch), yāu-gwo

*yautia* (Sp), yau-ti-a
*yawarakai rōru pan* (Jp), ya-wa-ra-ka-i rō-ru pan
*yeasts* (US), yeasts
*yé-dz* (Ch), yé-dz
*yegbintz* (Tr), yeg-hintz
*yēji* (Ch), yējī
*yellowtail snapper* (US), yel-low-tail snap-per
*yema* (Sp), yem-a
*yemiser selatta* (Af), ye-mi-ser se-lat-ta
*yemitas de mi bisabuela* (Mx), ye-mi-tas de mi bi-sa-bu-e-la
*yengeç* (Tr), yen-geç
*yepvadz* (Tr), yep-vadz
*yerakot* (Jw-Israel), ye-ra-kot
*yerba maté* (Sp), yer-ba ma-té
*yeşil salata* (Tr), ye-şil sa-la-ta
*yě-wèi* (Ch), yě-wèi
*yiaourti* (Gk), yia-our-ti
*yì-bèi* (Ch), yì-bèi
*yīng-táu* (Ch), yīng-táu
*yin-lyàu* (Ch), yin-lyàu
*yiouvetsi* (Gk), yiou-vet-si
*yodo* (Sp), yod-o
*yōgashi* (Jp), yō-ga-shi
*yogbourt* (Sw), yogh-ourt
*yogburt* (Da, Du, It), yogh-urt
*yogur* (Sp), yo-gur
*yogurt* (US), yo-gurt
*yoğurt* (Tr), yo-ğurt
*yogurtlu* (Tr), yo-gurt-lu
*yoğurtlu paça* (Tr), yo-ğurt-lu pa-ça

*yogurt salçasi* (Tr), yo-gurt sal-ça-si
*yoğurt tatlısı* (Tr), yo-ğurt tat-lı-sı
*yōkan* (Jp), yō-kan
*yōniku* (Jp), yō-ni-ku
*Yorkshire curd tart* (GB), York-shire curd tart
*Yorkshire pudding* (GB), York-shire pud-ding
*yosenabe* (Jp), yo-se-na-be
*yóu* (Ch), yóu
*yòu-dz* (Ch), yòu-dz
*yougburt* (Nw), yough-urt
*yóu-mài-yàn* (Ch), yóu-mài-yàn
*youngberry* (US), young-ber-ry
*you-tsài* (Ch), you-tsài
*yóu-yú* (Ch), yóu-yú
*yú* (Ch), yú
*yuba* (Jp), yu-ba
*yuca* (US), yuc-a
*yucatico* (Mx), yu-ca-ti-co
*yú-chr* (Ch), yú-chr
*yudeta* (Jp), yu-de-ta
*yude tamago* (Jp), yu-de ta-ma-go
*yufka* (Tr), yuf-ka
*yukka* (Tr), yuk-ka
*yukkai jang kuk* (Kr), yuk-kai jang kuk
*Yule log* (GB), Yule log
*yù-mi* (Ch), yù-mi
*yù-mi-myàr* (Ch), yù-mi-myàr
*yumurta başlama* (Tr), yu-mur-ta haş-la-ma
*yumurta lop* (Tr), yu-mur-ta lop

**yumurta rafadan** (Tr), yu-
mur-ta ra-fa-dan
**yún er** (Ch), yún er

**yusafandi** (Ar), yu-sa-fan-di
**yuzu** (Jp), yu-zu
**ywán-shwēi** (Ch), ywán-shwēi

# Z

**zabady** (Ar-Egypt), za-ba-dy
**zabaglione** (It), za-ba-glio-
ne
**zabibu** (Af-Swahili), za-bi-bu
**zacht gekookte eieren** (Du),
zacht ge-kook-te ei-e-ren
**začini** (SC), za-či-ni
**zadělavané žaludky** (Cz),
za-dě-la-va-né ža-lud-ky
**zaděnky** (Cz), za-děn-ky
**zafferano** (It), zaffe-ran-o
**zahter** (Ar), zah-ter
**zahuštěná** (Cz), za-hu-ště-ná
**zajac** (Po), za-jac
**zajíc** (Cz), za-jíc
**zajíc na smetaně** (Cz), za-jíc
na sme-ta-ně
**zakhari** (Gk), za-kha-ri
**zakoussotchnyï** (Rs), za-
kous-so-tchn-yï
**zakuski** (Rs), za-kus-ki
**zalivnoye iz rybi** (Rs), za-liv-
no-ye iz ry-bi
**zalm** (Du), zalm
**zamikand** (Ia), za-mi-kand
**zampe di maiale** (It), zampe di ma-ia-le

**zampone** (It), zam-po-ne
**zampone al cedro** (It), zam-
po-ne al ce-dro
**zanahorias** (Sp), za-na-ho-
ri-as
**Zander** (Gr), Zan-der
**Zander mit Mandeln** (Gr),
Zan-der mit Man-deln
**zanjabiyl** (Ar), zan-ja-biyl
**zapiekanka** (Po), za-pie-kan-
ka
**zapote** (Sp), za-po-te
**zarda** (Ia), zar-da
**zarigani** (Jp), zar-i-gan-i
**zarusoba** (Jp), zar-u-so-ba
**zarzamora** (Sp), zar-za-mo-
ra
**zarzuela** (Sp), zar-zue-la
**zarzuela de mariscos** (Sp),
zar-zue-la de ma-ris-cos
**zatar** (Ar), za-tar
**zatziki** (Gk), zat-zi-ki
**zavináč** (Cz), za-vi-náč
**zavyvanets** (Rs), za-vy-va-nets
**zayetz zharini v suharyakh**
(Rs), za-yetz zha-ri-ni v su-
har-yakh

**zayt** (Ar), zayt

**zayteem** (Jw-Israel), zay-teem

**zbeeb** (Ar), zbeeb

**z cytryna** (Po), z cyt-ry-na

**zdoba** (Rs), zdo-ba

**zedoary** (US), zed-o-a-ry

**zeera** (Ia), zee-ra

**zeevis** (Du), zee-vis

**zejtin** (SC), zej-tin

**zelena salata** (SC), ze-le-na sa-la-ta

**zelená paprika** (Cz), ze-le-ná pa-pri-ka

**zelen fasul** (Bu), ze-len fa-sul

**zeleninová polévka** (Cz), ze-le-ni-no-vá po-lév-ka

**zèleva chorba** (Bu), zè-le-va chor-ba

**zeli** (Cz), ze-li

**zeljanica** (SC), ze-lja-ni-ca

**zeller** (Hu), zel-ler

**zellerkrémleves** (Hu), zel-ler-krém-le-ves

**zellersaláta** (Hu), zel-ler-sa-lá-ta

**zemičke** (SC), ze-mič-ke

**žemle** (Cz), žem-le

**zenmai** (Jp), zen-mai

**zensai** (Jp), zen-sai

**zenzero** (It), zen-ze-ro

**zephyr** (US), zeph-yr

**zeppole** (It), zep-po-le

**žervé** (Cz), žer-vé

**ze śmietanką** (Po), ze ś-mie-tan-ką

**zest** (US), zest

**zesto** (Gk), zes-to

**zeytin** (Tr), zey-tin

**zeytinyăği** (Tr), zey-tin-yăği

**zeytinyăği sebzeter** (Tr), zey-tin-yăği seb-ze-ter

**zeytinyăği pirasa** (Tr), zey-tin-yăği pi-ra-sa

**zhá** (Ch), zhá

**zhá dà xīa** (Ch), zhá dà xīa

**zhá gèzi** (Ch), zhá gèzi

**zhāng chá yā** (Ch), zhāng chá yā

**zharennyi porosenok** (Rs), zha-ren-nyi po-ro-se-nok

**zharini orgutzi** (Rs), zha-ri-ni or-gu-tzi

**zhá yú tíaor** (Ch), zhá yú tíaor

**zhēng** (Ch), zhēng

**zhou fàn** (Ch), zhou fàn

**zhug** (Jw), zhug

**zibärtle** (Gr), zi-bär-tle

**zibda** (Ar), zib-da

**Ziegen** (Gr), Zie-gen

**zielone oliwki** (Po), zie-lo-ne o-liw-ki

**ziemniaki** (Po), ziem-nia-ki

**ziemniaki pure** (Po), ziem-nia-ki pu-re

**ziemniaki smazone** (Po), ziem-nia-ki sma-zo-ne

**Zigeuner Art** (Gr), Zi-geu-ner Art

**Zigeunerspiess** (Gr), Zi-geu-ner-spiess

**zik de venado** (Mx), zik de ve-na-do

**ziminu** (Fr), zi-mi-nu

**Zimt** (Gr), Zimt

**zinc** (US), zinc

**Zinfandel** (US), Zin-fan-del

**zingara** (Fr), zin-ga-ra

**Zink** (Gr), Zink

**ziti** (It), zi-ti

**ziti mezze** (It), zi-ti mez-ze

**ziti tagliati** (It), zi-ti ta-glia-ti

**žitne pabuljice** (SC), žit-ne pa-hu-lji-ce

**žitný chlèb** (Cz), žit-ný chléb

**zitoni** (It), zi-to-ni

**Zitrone** (Gr), Zi-tro-ne

**Zitronenschale** (Gr), Zi-tro-nen-scha-le

**zmrzlina** (Cz), zmr-zli-na

**zoet** (Du), zoet

**zöldbab** (Hu), zöld-bab

**zöldbableves** (Hu), zöld-bab-le-ves

**zöldbabsaláta** (Hu), zöld-bab-sa-lá-ta

**zöldpaprika** (Hu), zöld-pap-ri-ka

**zöldségleves** (Hu), zöld-ség-le-ves

**zöldségsaláta** (Hu), zöld-ség-sa-lá-ta

**zoni** (Jp), zo-ni

**zosui** (Jp), zo-sui

**zout** (Du), zout

**zraziki w sosie** (Po), zra-zi-ki w so-sie

**zrazy** (Po), zra-zy

**zrazy baranie** (Po), zra-zy ba-ra-nie

**zsemlegombóc** (Hu), zsem-le-gom-bóc

**zsemlék** (Hu), zsem-lék

**zsendice** (Hu), zsen-di-ce

**zsirban sült krumpli** (Hu), zsir-ban sült krump-li

**zsiványpecsenye** (Hu), zsi-vány-pe-cse-nye

**zucca** (It), zuc-ca

**zuccata** (It), zuc-ca-ta

**zucche ripiene** (It), zuc-che ri-pie-ne

**zucchero** (It), zuc-che-ro

**zucchini** (It), zuc-chi-ni

**zucchini fritti** (It), zuc-chi-ni frit-ti

**zucchini ripieni parizzi** (It), zuc-chi-ni ri-pie-ni pa-riz-zi

**zuccotto** (It), zuc-cot-to

**Zucker** (Gr), Zuc-ker

**Zuckererbsen** (Gr), Zuc-ker-erb-sen

**Zuger Kirschtorte** (Gr), Zu-ger Kirsch-tor-te

**Zuger Rötel** (Gr), Zu-ger Rö-tel

**Zunge** (Gr), Zun-ge

**Zungenwürst** (Gr), Zun-gen-würst

**zupa grzybowa** (Po), zu-pa gr-zy-bo-wa

**zupa jarzynowa** (Po), zu-pa jar-zy-no-wa

**zupa ogorkowa** (Po), zu-pa o-gor-ko-wa

**zupa owocowa** (Po), zu-pa o-wo-co-wa

**zupa ze sliwek** (Po), zu-pa ze sli-wek

**zuppa** (It), zup-pa

**zuppa alla pavese** (It), zup-pa al-la pa-ve-se

**zuppa di castagne** (It), zup-pa di cas-tagne

**zuppa di ceci** (It), zup-pa di ce-ci

**zuppa di frutti di mare** (It), zup-pa di frut-ti di mare

**zuppa di pesce** (It), zup-pa di pes-ce

**zuppa di verdura** (It), zup-pa di ver-du-ra

**zuppa Inglese** (It), zup-pa In-gle-se

**zuppa rustica** (It), zup-pa ru-sti-ca

**zur** (Po), zur

**zura** (Ar-Egypt), zu-ra

**Zürcher Leberspiessli** (Gr), Zür-cher Le-ber-spiess-li

**zushi** (Jp), zu-shi

**žuta boranija** (SC), žu-ta bo-ra-ni-ja

**Zutaten** (Gr), Zu-ta-ten

**zuur** (Du), zuur

**zuurkool** (Du), zuurkool

**Zwetschgen** (Gr-Austria), Zwetsch-gen

**Zwetschgenknödeln** (Gr), Zwetsch-gen-knö-deln

**Zwieback** (Gr), Zwie-back

**Zwiebelfleisch** (Gr), Zwie-bel-fleisch

**Zwiebeln** (Gr), Zwie-beln

**Zwiebelrostbraten** (Gr), Zwie-bel-rost-bra-ten

**Zwiebelsuppe** (Gr), Zwie-bel-sup-pe

**żywność** (Po), żyw-ność